A. Schaffranek

Floral Almanac

containing the flowering season of one thousand and seven hundred

phaenogamous plants of Florida

A. Schaffranek

Floral Almanac
containing the flowering season of one thousand and seven hundred
phaenogamous plants of Florida

ISBN/EAN: 9783337272401

Printed in Europe, USA, Canada, Australia, Japan

Cover: Foto ©berggeist007 / pixelio.de

More available books at **www.hansebooks.com**

FLORAL CALENDAR

CONTAINING THE

FLOWERING SEASON

OF

ONE THOUSAND AND SIXTY HUNDRED

PHAENOGAMOUS PLANTS OF FLORIDA.

BY

Dr. A. SCHAFFRANEK. Botanist.

PALATKA, FLORIDA

PRINTED BY THE
PALATKA NEWS PUBLISHING COMPANY
PALATKA, FLA.
1888

Mr. H. Fisher,

Dr GEO. VASEY

U. S. BOTANIST, DEPARTMENT OF AGRICULTURE, WASHINGTON, D. C.,

AND

Mr WILLIAM SAUNDERS

SUPERINTENDENT OF EXPERIMENT. GARDEN, DEPARTMENT OF AGRICULTURE.

WASHINGTON, D. C.,

Mr. P.   A.   D

WITH

LASTING ESTEEM AND FRIENDSHIP.

DR. A. SCHAFFRANEK, BOTANIST

PREFACE.

Having been requested since the last three years by Botanists in this and foreign countries to publish a list of our phaenogamous plants, and having heard so often the complaints of a great many visitors, that, while they spend the winter months in Florida, they could not find as many flowers as they expected, I thought it necessary by publishing this Floral Almanac to satisfy all of them at once. It has been condensed into the smallest compass in such an order and form as to facilitate as much as possible the collection of plants in their proper flowering season, thus enabling every botanist, agriculturist, florist and dendrologist to make a satisfactory selection for personal use and observation, and to prove to every visitor that nature has provided for us flowers all the year round. This Almanac will be not only a guide to any scientific visitor in the State, but may be used at the same time as a check-list.

If the arduous labor bestowed for years upon this work should render efficient aid to those who are earnestly interested in the Flora of Florida the author will be richly repaid for all his exertions.

DR. A. SCHAFFRANEK.

PALATKA. FLORIDA.

FLORAL ALMANAC.

MONTH.	No.	NAME OF PLANT.	NATURAL ORDER.	LOCALITY.
Jan'y to Feb'y.	1	Ulmus alata. Michx.	Ulmaceae.	Rich soil.
Jan'y to March.	2	Alnus serrulata. Ait.	Betulaceae.	River banks. Palatka.
" "	3	Andromeda phillyreaefolia. Hook.	Ericaceae.	Shallow ponds in pine barrens.
" "	4	Calamintha canescens. T.& G.	Labiatae.	Dry sand. West coast.
Feb'y to March.	5	Hepatica triloba. Chaix.	Ranunculaceae.	Shady woods. N. W. Florida.
" "	6	Nasturtium sessiliflorum. Nutt	Cruciferae.	Banks of the Chipola river.
" "	7	Dentaria lacinata. Muhl.	"	Riv. b'ks. shady spots. Palatka.
" "	8	Acer dasycarpum. Ehrh.	Aceraceae.	Riv. banks. Rice creek. Palatka.
" "	9	" rubrum. Lin.	"	Swamps.
" "	10	Cercis Canadensis. Lin.	Leguminosae.	Rich soil. Palatka.
" "	11	Cassia angustisiliqua. Lam.	"	Punta Gorda. South Florida.
	12	" biflora. Lin.	"	South Florida. Key West.
	13	Prunus umbellata. Ell.	Rosaceae.	Dry, light soil. Palatka.
	14	" Caroliniana.	"	River banks. Palatka.
	15	Amelanchier Canadensis. Lin.		Woods around Palatka.
	16	" Canad. var. Botryapium. Torr. & Gray.		Woods.
	17	Oldenlandia coerulea. Gray.	Rubiaceae.	Moist banks. Palatka.
	18	" rotundifolia. Gray.	"	Sandy soil near the coast.
	19	Fedia radiata. Michx.	Valerianaceae.	River banks. Common.
	20	Krigia Carolinea. Nutt.	Compositae.	Dry, sandy places.
	21	Leucothoë axillaris. Don.	Ericaceae.	Dry, sandy soil.
	22	Epigaea repens. Lin.	"	Sandy banks and soil.
	23	Vaccinium corymbosum. Lin.	"	Margins p'ds. swamps. Palatka.
	24	Benzoin odoriferum. Nees.	Lauraceae.	Banks of streams; low woods.
	25	" mellissaefolium. Nees.		Margins of ponds.
	26	Tetranthera geniculata. Nees.	"	Shallow pine barren ponds.
	27	Dirca palustris. Lin.	Thymeleaceae.	Shady banks of streams.
	28	Pachysandra procumbens. Michx.	Euphorbiaceae.	West Florida. Common.
	29	Ulmus fulva. Michx.	Ulmaceae.	Rich woods. West Florida.
	30	" Floridana. n. sp.	"	Banks of Chipola. Mariana rs.
	31	" Americana. Lin.	"	Low grounds. West Florida.
	32	Planera aquatica. Genel.	"	River swamps. Palatka.
	33	Corylus Americana. Wall.	Cupuliferae.	Rich soil. margins of woods.
	34	Myrica inodora. Bart.	Myricaceae.	Margins of swamps near coast.
	35	Leitneria Floridana.	"	Salt or brackish marshes. Ap'la.
	36	Taxodium distichum. Rich.	Coniferae.	Swamps and ponds.
	37	Corallorhiza odontorhiza. Nutt	Orchidaceae.	Shady woods.
	38	Trillium discolor. Wrang.	Smilaceae.	Rich woods in Middle Florida.
	39	Erythronium Americanum. Smith.	Liliaceae.	Rich woods in Middle Florida.

MONTH.	No.	NAME OF PLANT.	NATURAL ORDER.	LOCALITY.
Feb'y to April.	40	Chrysogonum Virginianum. L.	Compositae.	Dry, open woods.
" "	41	Chaptalia tomentosa. Vent.	"	Low pine barrens.
" "	42	Bartonia verna. Muhl.	Gentianaceae.	Damp pine barrens near coast.
March to April.	43	Zanthorhiza apiifolia. L'Her.	Ranunculaceae.	Shady banks.
" "	44	Asimina triloba. Dunal.	Anonaceae.	River banks. West Florida.
" "	45	" parviflora. Dunal.	"	Rich woods; Duval Co.: Palatka.
" "	46	" grandiflora. Dunal.	"	Sandy pine barrens; Put. Co.
"	47	Sanguinaria Canadensis Lin.	Papaveraceae.	Rich woods. Leon Co.
"	48	Corydalis aurea. Willd.	Fumariaceae.	Banks of Apalachicola river.
"	49	Nasturtium tanacetifolium. Hook.	Cruciferae.	Damp soil. East Florida.
	50	Cardamine Ludoviciana. Hook		Waste places near dwellings.
	51	" hirsuta. Lin.		Wet soil.
	52	Sisymbrium canescens. Nutt.	"	Waste grounds.
	53	Draba cuneifolia. Nutt.	"	West Florida.
	54	Senebiera Coronopus. Poir.	"	Waste places.
	55	Capsella Bursa-pastoris. Munch	"	Waste places.
	56	Viola villosa. Walt.	Violaceae.	Dry sandy or gravelly soil.
	57	Helianthemum Carolinianum. Mich.	Cistaceae.	Dry sandy soil.
	58	Helianthemum arenicola sp. n.	"	Drifting sands near the coast.
	59	Ascyrum pumilum. Mich.	Hypericaceae.	Dry gravelly soil.
	60	" microsepalum. T. & G.	"	Flat pine barrens.
	61	Stellaria prostrata. Baldw.	Caryophyllaceae.	Damp places.
	62	Geranium Carolinianum. Lin.	Geraniaceae.	Waste places. Common.
	63	Rhus aromatica. Ait.	Anacardiaceae.	Dry open woods. West Florida.
	64	Acer saccharinum var. Floridianum. Chap.	Aceraceae.	Upland woods. Middle Fla.
	65	Negundo aceroides. Mich.		River banks.
	66	Tirfolium Carolinianum. Mich	Leguminosae.	Waste places.
	67	Piscidia Erythrina. Lin.	"	South Florida.
	68	Baptisia leucantha. Torr. & Gray	"	River banks.
	69	Prunus Americana. Marsh.	Rosaceae.	Woods. Common.
	70	" Chicasa. Mich.	"	Old fields forming thickets
	71	Fragaria Virginiana. Ehrh.		Rich woods.
	72	Crataegus apiifolia. Mich.		River swamps.
	73	Crataegus arborescens. Ell.		River banks.
	74	" aestivalis. T. & G.		Margins of pine barrens.
	75	Pyrus arbutifolia. Lin.		Swamps. Putnam county.
	76	" arb. var. erythrocarpa. Chapm.		Swamps. Putnam county.
	77	Pyrus arbut. var. melanocarpa. Chapm.		Swamps. Putnam county.
	78	Fothergilla alnifolia. Lin.	Hamamelidaceae.	Swamps.
	79	Liquidambar styraciflua. Lin.	"	Swamps.
	80	Chaerophyllum Teinturieri. Hoz Arn.	Umbelliferae.	Banks of the Apalachicola river.
" "	81	Ernodea littoralis. Swartz.	Rubiaceae.	Along the coast. South Florida.
" "	82	Mitchella repens. Lin.	"	Shady woods. Palatka.
" "	83	Morinda Roioes. Lin.	"	South Florida.
" "	84	Gelsemium sempervirens.	"	Margins of swamps, Palatka.
" "	85	Erigeron vernum. T. & G.	Compositae.	Pine barren swamps.
" "	86	Senecio lobatus. Pers.	"	Low grounds.
" "	87	Vaccinium nitidum. Andr.	Ericaceae.	Low pine barrens.
" "	88	" myrsinites. Mich.	"	Sandy pine barrens.
" "	89	" tenellum.	"	Margins of pine barrens.

MONTH.	No.	NAME OF PLANT.	NATURAL ORDER	LOCALITY.
March to April.	146	Betula nigra. Lin.	Betulaceae.	Banks of rivers.
" "	147	" rubra. Lin.	"	Banks of rivers.
" "	148	Salix rosmarinifolia.	Salicaceae.	Swamps and low grounds.
" "	149	" Floridiana n. sp.	"	Rocky banks. West Florida.
" "	150	" nigra. Marshall.	"	Swamps and muddy river banks
" "	151	" angulata. Ait.	"	River banks.
" "	152	Pinus serotina. Mich.	Coniferae	Borders swamps; lower districts
" "	153	" Taeda. Lin.	"	Light mostly damp soil. Palatka
" "	154	" australis. Mich.	"	Sandy soil. Palatka.
" "	155	Juniperus Virginiana. Lin.	"	Dry, rocky or even wet soil.
" "	156	Cupressus thyoides. Lin.	"	Swamps. Palatka.
" "	157	Taxus Floridiana. Nutt.	"	Banks of Apalachicola river.
" "	158	Torreya taxifolia. Arn.	"	Low grounds. Murphy Island.
" "	159	Zamia integrifolia. Willd.	Cycadaceae.	Rich woods.
" "	160	Arisaema Dracontium. Schott.	Araceae.	Ponds and slow flowing streams.
" "	161	Orontium aquaticum. Lin.	"	Wet pine barrens. Palatka.
" "	162	Calopogon parviflorus. Lindl.	Orchidaceae.	Wet pine barrens. Palatka.
" "	163	" multiflorus. Lindl.	"	Rich damp soil.
" "	164	Amaryllis Atamasco. Lin.	Amaryllidaceae	Low ground. Palatka suburbs.
" "	165	Hypoxis erecta. Lin.	"	Low pine barrens. Palatka.
" "	166	Hypoxis juncea. Smith.	"	Woods and thickets.
" "	167	Smilax Pseudo China. Lin.	Smilaceae.	Pine barren ponds and swamps.
" "	168	" Walteri. Pursh.	"	Woods and thickets.
" "	169	Uvularia perfoliata. Lin.	Melanthaceae.	Low shady woods. Middle Fla.
" "	170	" Floridiana. n. sp.	"	Swamps and bogs.
" "	171	Carex Bromoides. Schn.	"	Swamps. West Florida.
" "	172	Trisetum palustre. Torr.	Gramineae.	Swamps. West Florida.
" "	173	Viburnum densiflorum.	Caprifoliaceae	Wooded hillsides. Palatka.
" "	174	" nidum var. angustifolium. Gray & Torr.	"	Margins of swamps. Palatka.
	175	Richardsonia scabra. St. Hil.	Rubiaceae.	Margins of swamps. Palatka.
	176	Galium pilosum var. puncticulosum. Gray.	"	Dry, rich soil. Palatka suburbs.
" "	177	Oldenlandia patens.	"	Roadsides. Palatka suburbs.
" "	178	Catesbaea parviflora. Swartz.	"	Bahia Honda. South Florida.
" "	179	Vaccinium virgatum. Ait.	Ericaceae.	Low pine barrens.
April to May.	180	Thalictrum anemonoides. Mich	Ranunculaceae.	Woods.
" "	181	Ranunculus palmatus. Ell.	"	Swamps in the pine barrens.
" "	182	Isopyrum biternatum. T. & G.	"	Shady woods. West Florida.
" "	183	Aquilega Canadensis. Lin.	"	Rocky woods. West Florida.
" "	184	Magnolia grandiflora. Lin.	Magnoliaceae.	Light fertile soil. Palatka.
" "	185	Podophyllum peltatum. Lin.	Berberidaceae.	Rich woods.
" "	186	Sarracenia purpurea. Lin.	Sarraceniaceae.	Mossy swamps.
" "	187	" Psittacina. Mich.	"	Pine barren swamps.
" "	188	" Drummondii. Croom	"	Pine barren swamps.
" "	189	" flava. Lin.	"	Low pine barrens. Palatka.
" "	190	" variolaris. Mich.	"	Low pine barrens. Pal. suburbs.
" "	191	Argemone Mexicana. Lin.	Papaveraceae.	Waste places. Palatka.
" "	192	Cardamine rhomboides. D. C.	Cruciferae.	Cool springs. West Florida.
" "	193	Lepidium Virginicum. Lin.	"	Waste places. Palatka.
" "	194	Senebiera pinnatifida. D. C.	"	Waste places. Common.
" "	195	Viola lanceolata. Lin.	Violaceae.	Low pine barrens.
" "	196	" hastata. Mich.	"	Shaded hillsides.
" "	197	Helianthemum corymbosum. Mich.	Cistaceae.	Dry sandy soil.
" "	198	Helianthemum Canadense. M.		Dry sterile soil.

MONTH	No.	NAME OF PLANT.	NATURAL ORDER.	LOCALITY.
April to May.	199	Drosera filiformis. Raf.	Droseraceae.	Low pine barrens. Palatka.
"	200	" capillaris. Poir.	"	Boggy ponds. Apalachicola.
"	201	" brevifolia. Pursh.	"	Low sandy pine barrens.
"	202	Spergula arvensis. Lin.	Caryophllaceae.	Cultivated fields.
"	203	Alsine squarrosa. Fenzl.	"	Dry sandy hills. West Florida.
"	204	Arenaria serpyllifolia. Lin.	"	Waste places. Common.
"	205	Cerastium vulgatum. Lin.	"	Fields.
"	206	" viscosum. Lin.	"	Fields.
"	207	Silene Baldwinii. Nutt.	"	Low shady woods.
"	208	Stuartia Virginica. Car.	Camelliaceae.	Shady woods.
"	209	Oxalis stricta. Lin.	Oxalidaceae.	Rich woods. West Florida.
"	210	Ceanothus microphyllus. Mich	Rhamnaceae.	Dry barrens.
"	211	Evonymus Americana. Lin.	Celastraceae.	Low shady woods. Palatka.
	212	Aesculus Pavia. Lin.	Sapindaceae.	Rich soil.
	213	Polygala nana. D. C.	Polygalaceae.	Low shady pine bar'ns. Palatka
	214	Lupinus perennis. Lin.	Leguminosae.	Dry sandy soil.
	215	" villosus. Willd.	"	Dry sandy barrens. Palatka.
	216	" diffusus. Nutt.	"	Dry sandy ridges.
	217	Trifolium reflexum. Lin.	"	Waste places and pastures.
	218	Psoralea canescens. Mich.	"	Dry pine barrens.
	219	Astragalus glaber. Mich.	"	Dry pine barrens.
	220	" villosus. Mich.	"	Dry pine barrens.
	221	Vicia micrantha. Nutt.	"	Banks of rivers; shady places.
	222	" acutifolia. Ell.	"	Damp soil near the coast.
	223	Erythrina herbacea. Lin.	"	Light sandy soil. Palatka.
	224	Baptisia lanceolata. Ell.	"	Dry pine barrens.
	225	" alba. R. Brown.	"	Damp soil.
	226	Prunus serotina. Ehrh.	Rosaceae.	Woods. Common.
	227	Spiraea opulifolia. Lin.	"	Banks of streams. Palatka.
	228	Rubus villosus. Ait.	"	Swamp thickets. Palatka.
	229	" cuneifolius. Pursh.	"	Old fields. Common.
	230	" trivialis. Michx	"	Dry sandy soil. Palatka.
	231	Crataegus spathulata. Michx	"	River banks.
	232	" Crus-galli. Lin.	"	Woods.
	233	" coccinea. Lin	"	Open dry woods.
	234	" glandulosa. Michx	"	Dry pine barrens.
	235	" parvifolia. Ait.	"	Sandy soil.
	236	Pyrus angustifolia. Ait	"	Open woods.
	237	Calycanthus laevigatus. Willd	Calycanthaceae.	Banks of streams; low country.
	238	Philadelphus grandiflorus. Willd.	"	Banks of streams.
	239	Hydrocotyle umbellatus. Lin.	Umbelliferae.	Wet places. Palatka.
	240	Cornus stricta. Lin.	Cornaceae.	Swamps. Common.
	241	Nyssa aquatica. Lin.	"	Ponds and swamps. Palatka.
	242	" uniflora. Walt.	"	Deep swamps and ponds.
	243	" capitata. Walt.	"	Swamps near the coast.
	244	Viburnum prunifolium. Lin.	Caprifoliaceae.	Dry rich woods. Sub's Palatka.
	245	" obovatum. Walt.	"	River banks.
	246	" nudum. Lin.	"	Swamps. Common.
	247	" dentatum. Lin.	"	Rich damp soil. West Florida.
	248	Chrysopsis obligantha. Chap.	Compositae.	Low pine barrens. Palatka.
	249	Hymenopappus scabiosaeus. L'Her.	"	Light dry soil. Palatka.
	250	Leptopoda Helenium. Nutt.		Margins of pine barren ponds.
	251	" fimbriata. Torr. & Gray.		Low pine barren.

MONTH.	No.	NAME OF PLANT.	NATURAL ORDER	LOCALITY
April to May.	252	Leptopoda puberula. Macbride.	Compositae.	Wet pine barren.
" "	253	Marshallia lanceolata. Pursh.	"	Dry open woods.
" "	254	Antennaria plantaginifolia. Hook.		Sterile soil.
	255	Senecio tomentosus. Michx.		Damp soil.
	256	" Elliottii. Torr. & Gray.		Rocky places. West Florida
	257	Arnica nudicaulis. Ell.		Wet pine barren. Palatka.
	258	Cirsium horridulum. Michx.		Sandy soil.
	259	Apogon humilis. Ell.		All over Florida.
	260	Krigia Virginica. Willd.		Dry sandy soil.
	261	Cynthia Dandelion. D. C.		Damp soil.
	262	Lygodesmia aphylla. D. C.		Dry sandy pine barren.
	263	Pyrrhopappus Carolineanus. D. C.		Fields. Common.
	264	Gaylussacia frondosa. Torr. & Gray.	Ericaceae.	Low ground. Palatka.
	265	Gaylussacia dumosa. Torr & Gray.		Low ground. Palatka suburbs.
	266	Leucothoë racemosa. Gray.		Margins ponds and swamps near Palatka.
	267	Andromeda nitida. Bart.		Low pine barrens near Palatka.
	268	" Mariana. Lin.		Damp soil. Palatka and coast.
	269	" speciosa. Michx.		Low pine barrens. Palatka.
	270	" ferruginica. Walt.		Low sandy pine barrens.
	271	Oxydendron arboreum. D. C.		Rich woods. Red Water branch
	272	Rhododendron nudiflorum. Torr.		Swamps and banks of streams.
	273	Ilexopaca. Ait.	Aquifoliaceae.	Sandy soil. Palatka suburbs.
	274	" Dahoon. Walt.	"	Margins swamps and ponds.
	275	" var. myrtifolia. Walt.		Margins of swamps. Light sandy soil along the coast.
" "	276	Ilexopaca Cassine. Lin.	"	
" "	277	" decidua. Walt.	"	River swamps. Palatka.
" "	278	" monticola. Gray.	"	Sandy margins of swamps.
" "	279	" verticillata. Gray.	"	Low grounds. West Florida.
" "	280	Styrax pulverulentum. Michx.	Styracaceae.	Pine barren swamps.
" "	281	" grandifolium. Ait.	"	Rich woods.
" "	282	Plantago heterophylla.	Plantaginaceae.	Waste places and fields.
" "	283	Plumbago scandens. Lin.	Plumbaginaceae.	South Florida.
" "	284	Utricularia inflata. Walt.	Lentibulaceae.	Ponds and ditches. Palatka.
" "	285	Tecoma stans. Jurr.	Bignoniaceae.	South Florida.
" "	286	Conopholis Americana. Walt.	Orobanchaceae.	Shady woods. Palatka.
" "	287	Aphyllon uniflorum. Torr. & Gray.		Woods in the vicinity of Palatka
	288	Linaria Floridana. Nov. spec.	Scrophulariaceae.	Drifting sand near the coast.
	289	Gratiola Virginiana. Lin.	"	Muddy banks and ditches.
	290	" Floridana. Nutt.	"	Muddy banks of Chipola river.
	291	" sphaerocarpa. nov. s.	"	Springs and rivulets. Palatka.
	292	Dianthera crassifolia. nov. s.	Acanthaceae.	Wet pine barrens. Apalachicola
	293	Salvia lyrata. Lin.	Labiatae.	Sandy soil. Palatka suburbs.
	294	Lithospermum hirtum. Lehm.	Borraginaceae.	Dry pine barrens.
	295	Nemophila microcalyx. Fisch. & Meyer.	Borraginaceae.	Shady woods. Palatka.
	296	Phlox divaricata. Lin.		Woods and banks. Palatka.
	297	" Walteri. Gray.		Dry gravelly hills; pine barrens.
	298	" pilosa. Lin.		Dry woods. Palatka.

MONTH.	No.	NAME OF PLANT.	NATURAL ORDER.	LOCALITY.
April to May.	299	Phlox subulata.	Borraginaceae.	Sandy pine barrens.
" "	300	Limnanthemum trachyspermum. Gray.	Gentianaceae.	Ponds in deep water.
	301	Amsonia ciliata. Walt.	Apocynaceae.	Dry sandy soil.
	302	Asclepias Curassavica. Lin.	Asclepiadaceae.	South Florida.
	303	" amplexicaulis. Michx.		Dry sandy pine barrens.
	304	" Michauxii. Decaisn.	"	Low sandy pine barrens.
	305	Chionanthus Virginica. Lin.	Oleaceae.	Light soil.
	306	Brunnichia cirrhosa. Banks.		River banks.
	307	Comandra umbellata. Nutt.	Santalaceae.	Dry soil.
	308	Phoradendron flavescens. Nutt.	Loranthaceae.	Damp soil. Palatka.
" "	309	Castanea pumila. Michx.	Cupuliferae.	Dry, sandy soil. Palatka.
" "	310	Fagus ferruginea. Ait.	"	Damp, sandy soil.
" "	311	Peltandra Virginica. Raf.	Araceae.	Marshes and wet places.
" "	312	Acorus Calamus. Lin.	"	Wet places.
" "	313	Epidendron venosum. Lindl.	Orchidaceae.	South Florida, on trees.
" "	314	Pogonia ophioglossoides. Nutt	"	Swamps. Palatka.
" "	315	Spiranthes gracilis. Bigel.	"	Damp soil. Palatka.
" "	316	Amaryllis Treatiae. Wats.	Amaryllidaceae.	Low ground, between Palatka and Rice Creek.
	317	Pancratium rotatum. Ker.	"	Low banks. Palatka.
	318	Tillandsia caespitosa. Lecoute.	Bromeliaceae.	East Florida, in large roundish clusters on trunks of trees.
	319	Iris versicolor. Lin.	Iridaceae.	Wet places. Palatka.
	320	" hexagona. Walt.	"	Swamps. Palatka.
	321	" cuprea. Pursh.	"	Swamps. East Palatka.
	322	Croomia pauciflora. Torr.	Roxburghiaceae.	Shady woods.
	323	Veratrum intermedium. n. sp.	Melanthaceae.	Rich shady hammocks. Palatka
	324	Schoenocaulon gracilis. Gray	"	Dry sands.
	325	Juncus maritimus. Lam.	Juncaceae.	Brackish marshes along coast.
	326	Tradescantia Virginica. Lin.	Commelinaceae.	Dry sandy soil. Palatka.
	327	Xyris brevifolia. Michx.	Xyridaceae.	Low sandy pine barns. Palatka
	328	" flabelliformis. n. sp.	"	Low pine barrens near the coast.
	329	Eriocaulon flavidulum. Michx.	Eriocaulonaceae.	Low sandy pine barrens.
	330	Carex stipata. Muhl.	Cyperaceae.	Swamps. Common.
	331	" Steudelii. Keenth.		Woody hillsides.
	332	" cephalophora. Muhl.	"	Dry soil. Palatka.
	333	" retroflexa. Muhl.		Open woods.
	334	" stellulata. Good.		Shady river swamps. Palatka.
	335	" sterilis. Willd.		Shady river swamps. Palatka.
	336	" conferta.		Pine barren swamps.
	337	" fcenea. Muhl.		Marshes.
	338	" polytrichoides.		Bogs and swamps. Palatka.
	339	" hirsuta.		Damp soil.
	340	" oxylepis. Torr. & Hook.		Low ground.
" "	341	" dasycarpa. Muhl.		Shady woods. Palatka.
" "	342	" lucosum. Willd.		Dry sandy soil.
" "	143	" nigromarginata. Schm.		Dry sandy soil.
" "	344	" grisea. Wahl.		Low ground.
" "	345	" Emmonsii. Dew.		Dry sandy soil.
" "	346	" granularis. Muhl.		Meadows and banks of streams.
" "	347	" striatula. Michx.		Dry open woods; margins fields.
" "	348	" styloflexa. Buckl.		Shady swamps. Palatka.
" "	349	" digitalis. Willd.		Low grounds.
" "	350	" lupalina. Muhl.		Deep river swamps. Palatka.

MONTH.	No.	NAME OF PLANT.	NATURAL ORDER.	LOCALITY
March to April.	351	Carex gigantea. Rudge.	Cyperaceae.	Pine barren ponds.
" "	352	" Baltzellii. Chapm.	"	Dry sandy soil. Middle Florida.
" "	353	" denticulata. Muhl.	"	Meadows and low ground.
" "	354	" venusta. Dew.	"	Low banks streams and swamps
" "	355	" verrucosa. Ell.	"	Margins and ponds of rivers.
" "	356	" Cherokeensis. Schk.	"	Banks of Apalachicola river.
" "	357	" debilis. Dew.	"	Low banks streams and swamps
" "	358	" glaucescens. Ell.	"	Pine barren ponds. Palatka.
" "	359	" decomposita.	"	Wet margins ponds and streams
" "	360	" bromoides. Schk.	"	Swamps and bogs.
" "	361	" crus corvi. Muhl.	"	River swamps. Palatka.
" "	362	" folliculata. Lin.	"	Wet margins of streams.
" "	363	" Halei. Carey.	"	Banks of Apalachicola river.
" "	364	" intumescens. Rudge.	"	Shady swamps.
" "	365	" Elliottii. Schu. & Torr.	"	Boggy margins p. br'n streams.
" "	366	" riparia. Curt.	"	Deep marshes.
" "	367	" striata. Michx.	"	Pine barren swamps. Palatka.
" "	368	" turgescens. Torr.	"	Pine barren swamps.
" "	369	" gymandra. Schn.	"	Damp woods.
" "	370	Sporobolus junceus. Keenth.	Gramineae.	Dry pine barren. Common.
" "	371	Stipa avenacea. Lin.	"	Dry soil.
" "	372	Streptachne Floridana. n. sp.	"	South Florida.
" "	373	Eatonia obtusata. Gray.		Dry soil.
" "	374	Melica mutica. Walt.		Dry open woods.
" "	375	Poa cristata. Walt.		Dry soil near Quincy. Mid. Fla.
" "	376	Festuca myurus. Lin.		Dry sandy soil.
" "	377	" tenella. Willd.		Dry sandy soil.
" "	378	" durinscula. Lin.		Common around dwellings.
" "	379	Phalaris intermedia. Bosc.		Shady places along coast. St. A.
" "	380	Thurberia Arkansana. Benth.		Shell mound near Apalachicola.
" "	381	Poa brevifolia. Muhl.		Rich soil around Palatka.
May to June.	382	Clematis Baldwinii. Torr. & Gray.	Ranunculaceae.	South Florida.
	383	Clematis reticulata. Walt.	"	Dry sandy soil.
	384	" Catesbyana. Pursh.		Dry sandy soil near the coast.
	385	Delphinium azureum. Michx.		Rich soil. West Florida.
	386	Illicium Floridanum. Ellis.	Magnoliaceae.	Sandy swamps.
	387	" parviflorum. Michx.	"	East Florida.
	388	Schizandra coccinea. Michx.		Shady woods. Palatka to Rieve.
	389	Magnolia glauca. Lin.		Swamps. Palatka.
	390	" Fraseri. Walt.		Rich woods. Northwest Fla.
	391	" cordata. Michx.		Rich slanted soil. Palatka.
	392	Liriodendron Tulipifera. Lin.		Low ground. N. and NW. Fla.
	393	Asimina pygmaea. Dunal.	Anonaceae.	
	394	" reticulata. Shuttlew.	"	South Fla. Mangrove swamps.
	395	Anona laurifolia. Dunal.		Banks Caloosa and Miami rivers
	396	Calycocarpum Lyoni. Nutt.	Menispermaceae.	Banks of Apalachicola river.
	397	Nymphaea odorata. Ait.	Nymphaeaceae.	Ponds and still water. St. Johns
	398	" flava. Leitner.		Ponds and still water. Rice cr'k.
	399	Cakile maritima. Scop.	Cruciferae.	Keys of South Florida.
	400	Polanisia tenuifolia. T. & G.	Capparidaceae.	South Florida.
	401	Cleome pungens. Willd.	"	Waste places.
	402	Gynandropsis pentaphylla. D. C.		Waste places.
	403	Capparis Jamaicensis. Jacqu.		South Florida.
	404	" cynophallaphora. Lin.		South Florida.

MONTH	No.	NAME OF PLANT.	NATURAL ORDER.	LOCALITY.
May to June.	105	Drosera longifolia. Raf.	Droseraceae.	Sandy swamps; often in water.
" "	106	" rotundifolia. L.		Mossy swamps Red & W. wat br.
" "	107	Ascyrum amplexicaule. Michx	Hypericaceae.	Damp soil near coast.
" "	108	Hypericum myrtifolium. Lam.	"	Pine barren ponds. Palatka.
" "	109	Silene Antirrhina. Lin.	Caryophyllaceae.	Low shady woods. Palatka.
" "	110	Oxalis violacea. Lin.	Oxalidaceae.	Rich woods. West Florida.
" "	111	Ptelea trifoliata. Lin.	Rutaceae.	Rocky banks.
" "	112	" Baldwinii. Torr. & G.	"	East Florida.
" "	113	Vitis cordifolia. Michx.	Vitaceae.	River swamps around Palatka.
" "	114	Euonymus atropurpurens. Jacqu.	Celastraceae.	River banks near Palatka.
	115	Cardiospermum Halicacubum. Lin.	Sapindaceae.	South Florida.
	116	Crotalaria Purshii. D. C.	Leguminosae.	Flat grassy pine barrens.
	117	Medicago lupulina. Lin.	"	Waste places. Palatka.
	118	Psoralea melilothioides. Michx		Dry soil.
	119	" Lupinellus. Michx.		Dry pine barrens.
	120	Amorpha fruticosa. Lin.		River banks. Palatka.
	121	Wistaria frutescens. D. C.		Margins of swamps. Palatka
	122	Astragalus obcordatus. Ell.		Dry sandy pine barrens.
	123	Rhynchosia galactoides. Nutt.		Dry sandy ridges.
	124	Galactia Elliottii. Nutt.		Dry soil.
	425	Baptisia megacarpa. Chap.		Light rich soil. Middle Florida.
	126	" Lecontei. Torr. & G.		Dry sandy soil.
	127	Chrysobalanus oblongifolius. Michx.	Rosaceae.	Dry sandy pine barrens. Palatka
	128	Chrysobalanus Icaco. Lin.		Indian river. South Florida.
	129	Rosa lucida. Ehrh.	"	Mostly in dry soil.
	130	Crataegus flava. Ait.		Shady sandy places.
	431	Eugenia dichotoma. D. C.	Myrtaceae.	South Florida.
	432	" procera. Poir.		Indian river.
	433	" monticola. D. C.		South Florida.
	434	" buxifolia. D. C.		South Florida.
	435	Calyptranthus Chytraculia. Swartz.	"	South Florida.
	436	Oenothera linearis. Michx.	Onagraceae.	Dry light soil.
	437	Opuntia Ficus indica. Han.	Cactaceae.	South Florida.
	438	" polyantha. Han.	"	Palatka; waste places. Key W.
	439	Itea Virginica. Lin.	Saxifragaceae.	Swamps. Common.
	440	Hydrangea quercifolia.	"	Shady banks.
	441	Decumaria barbara. Lin.	"	Banks of streams. St. Johns.
	442	Hydrocotile ranunculoides. Lin	Umbelliferae.	Wet places. Palatka.
	443	Thaspium barbinode. Nutt.	"	River banks.
	444	" aureum. Nutt.		Rich soil.
	445	Cornus alternifolia. L'Her.	Cornaceae.	Banks of St. Johns. Palatka.
	446	" florida. Lin.	"	Along White Water. Palatka.
	447	" asperifolia. Michx.	"	Wet soil.
	448	" sericea. Lin.	"	Low woods. Palatka.
	449	" paniculata. L'Her.	"	North Florida.
	450	Nyssa multiflora. Wang.		Rich upland woods.
	451	Viburnum acerifolium. Lin.	Caprifoliaceae.	Dry open woods. West Florida.
	452	" scabrellum. T. & G.		Swamps.
	453	Randia aculeata. Lin.	Rubiaceae.	South Florida.
	454	Gardenia clusiaefolia. Jacqu.	"	South Florida.
	455	Pinckneya pubens. Michx.	"	Marshy banks streams. Palatka
	456	Spigelia gentianoides. Chapm.	"	Light, dry soil.

MONTH.	No.	NAME OF PLANT.	NATURAL ORDER.	LOCALITY.
May to June.	157	Spigelia Marilandica. Lin.	Rubiaceae.	Rich woods.
" "	158	Coreopsis grandiflora. Nutt.	Compositae	Dry soil around Palatka.
" "	159	" lanceolata. Lin.	"	Dry, rich soil.
" "	160	Leptopoda brachypoda. T. & G.		River banks.
" "	161	Maruta Cotula. D. C.		Introduced.
" "	162	Achillea millefolium. Lin.		Old fields and around dwellings.
" "	163	Leucanthemum vulgare. Lam.		Fields near Palatka.
" "	164	Vaccinium stamineum. Lin.	Ericaceae.	Dry woods around Palatka.
" "	165	" arboreum. Michx.	"	Open woods around Palatka.
" "	166	Andromeda ligustrina. Muhl.	"	Margins of swamps, Palatka.
" "	167	Kalmia latifolia. Lin.	"	Shady banks, Palatka.
" "	168	Ilex glabra. Gray.	Aquifoliaceae.	Low pine barrens.
" "	169	Prinos coriaceus. Ell.	"	Wet thickets.
" "	470	Diospyros Virginiana. Lin.	Ebenaceae.	Woods and old fields, Palatka.
" "	471	Ardisia Pickeringia. T. & G.	Myrsinaceae.	Halifax river.
" "	472	Samolus ebracteatus. Kunth.	Primulaceae.	Saline marshes.
" "	473	Utricularia fibrosa. Walt.	Lentibulaceae.	Ponds, ditches around Palatka.
" "	474	" purpurea. Walt.		Shallow ponds.
" "	475	Tecoma radicans. Juer.	Bignoniaceae.	Woods around Palatka.
" "	476	Catalpa bignonioides. Walt.	"	River banks.
" "	477	Schwalbea Americana. Lin.	Scrophulariaceae.	Sandy pine barrens.
" "	478	Brunella vulgaris. Lin.	Labiatae.	Low grounds.
" "	479	Onosmodium Virginianum. D. C.	Borraginaceae.	Dry pine barrens.
	480	Phlox Floridana. Benth.	Polemoniaceae.	Dry, open woods. Middle Fla.
	481	Cuscuta neuropetela. Engelm.	Convolvulaceae.	Field, sterile soil on small hsrbs.
	482	Amsonia Tabernaemontana. Walt.	Apocynaceae.	Swamps and wet banks.
	483	Vallesia chiococcroides. Kunth.	"	South Florida.
	484	Vinca rosea. Lin.	"	Around Apalachicola.
	485	Asclepias variegata. Lin.	Asclepiadaceae.	Dry, open woods; borders fields.
	486	Aristolochia tomentosa. Lin.	Aristolochiaceae.	River banks.
	487	Euphorbia inundata. Torr.	Euphorbiaceae.	Pine barren swamps.
	488	" telephioides. n. sp.		Low sandy p. barr'ns near coast.
	489	" Ipecacuanha. Lin.		Dry sandy soil.
	490	Xanthosoma sagittifolium. Schott.	Araceae.	Marshy, springy places, Palatka.
	491	Pistia spathulata. Michx.	"	In still water, St. Johns, Palatka
	492	Sagittaria pusilla. Pursch.	Alismaceae.	Shallow ponds and streams.
	493	Calopogon pallidus. n. sp.	Orchidaceae.	Wet pine barrens near thickets.
	494	Epidendron cochleatum. Lin.	"	South Florida.
	495	" umbellatum. Swartz.	"	South Florida. Miami.
	496	Polystachia luteola. Hook.	"	On trunks different trees. S. Fla.
	497	Dendrophylax Lindenii. Reichenb.	"	On Oreodoxa regia, Indian river.
	498	Habenaria Garberi. Porter.	"	Damp shady woods, Manatee co.
	499	Pogonia divaricata. R. Brow.	"	Swamps. Palatka.
	500	" verticillata. Nutt.	"	Low shady woods, Rice creek.
	501	Spiranthes tortilis. Willd.	"	Low or marshy p. bar'ns Palatka
	502	Aletris farinosa. Lin.	Haemadoraceae.	Ponds, ditches around Palatka.
	503	" aurea. Walt.	"	Wet pine barrens. Palatka to Brown's Landing.
" "	504	Tillandsia bulbosa. Hook.	Bromeliaceae.	South Florida.
" "	505	" juncea. Lecomte.	"	South Florida.
" "	506	" Bertramii. Ell.	"	South Florida.
" "	507	" recurvata. Pursh.	"	East Florida.

MONTH.	No.	NAME OF PLANT.	NATURAL ORDER.	LOCALITY.
May to June.	508	Nemastylis coelestina. Nutt.	Iridaceae.	Pine barrens around Palatka.
" "	509	Smilax glauca. Walt.	Smilaceae.	Shady margins of swamps.
	510	" auriculata. Walt.	"	Dry sandy ridges along coast.
	511	" rotundifolia. Lin.	"	Swamps. Common. Palatka.
	512	Caprosmanthus herbaceus. Kunth.		Dry, fertile soil. Palatka.
	513	Caprosmanthus peduncularis. Kunth.		Rich soil.
	514	Medeola Virginica. Lin.	"	Shady banks. Middle Florida.
	515	Polygonatum biflorum. Ell.	Liliaceae.	Shady banks. N. and W. Fla.
	516	Allium mutabile. Michx.	"	Dry, sandy soil.
	517	Schoenolirion Michauxii. Torr.	"	Pine barren swamps around Pal.
	518	Yucca gloriosa. Lin.	"	Drifting sands near the coast.
	519	" filamentosa. Lin.	"	Light or sandy soil. Palatka.
	520	" aloifolia. Lin.	"	Sandy soil along the coast.
	521	Amianthium muscaetoxicum. Gray.	Melanthaceae.	Rich woods.
	522	Amianthium angustifolium. Gray.		Low pine barrens.
	523	Chamaelirium luteum. Gray.	"	Low grounds, Palatka's vicinity
	524	Juncus tenuis. Willd.	Juncaceae.	Low grounds, common. Palatka
	525	" dichotomus. Ell.		Low grounds. Palatka.
	526	Lachnocaulon Michauxii. Kunth.	Eriocaulonaceae.	Boggy places around Palatka.
	527	Eleocharis pygmaea. Torr.	Cyperaceae.	Muddy or sandy b'nks near coast
	528	Zizania milliacea. Michx.	Gramineae.	Deep marshes and ponds.
	529	Alopecurus geniculatus. Lin.	"	Wet cultiv'ted ground, common
	530	Poa flexuosa. Muhl.	"	Rich, shaded soil. Palatka.
" "	531	" angustifolia. Ell.	"	Rich soil, mostly neardwellings.
" "	532	" compressa. Lin.	"	Rich soil, mostly near dwellings.
" "	533	Paspalum praecox. Walt.	"	Pine bar'n swamps near Palatka
" "	534	Panicum latifolium. Lin.	"	Dry, rich soil.
" "	535	" scoparium. Lin.	"	Open woods and margins fields.
" "	536	" divaricatum. Lin.	"	South Florida. Everglades.
" "	537	" viscidum. Ell.	"	Wet swamps; bogs near coast.
" "	538	" scabriusculum. Ell.	"	Pine barren swamps. Common.
May to July.	539	Malva rotundifolia. Lin.	Malvaceae.	Around dwellings.
" "	540	Polygala setacea. Michx.	Polygalaceae.	Low pine barrens.
" "	541	" polygama. Walt.	"	Wet or dry sandy pine barrens.
" "	542	" Boykinii. Nutt.	"	Rich calcareous soil.
" "	543	Crotolaria ovalis. Pursh.	Leguminosae.	Dry pine barrens.
" "	544	Rhynchosia tomentosa. var. volubilis. Gray & Torr.		Dry sandy soil.
	545	Samolus floribundus. Kunth.	Primulaceae.	Brackish marshes.
	546	Dianthera ovata, var. augusta. Wall.	Acanthaceae.	Pine barren ponds.
	547	Smilax tamnoides. Lin.	Smilaceae.	Swamps and thickets.
	548	Juncus setaceus. Ros. & K.	Juncaceae.	Low grounds and swamps.
	549	Dichromena leucocephala. Michx.	Cyperaceae.	Damp soil around Palatka.
May to August	550	Polygala verticillata. Lin.	Polygalaceae.	Dry sandy soil.
" "	551	" leptostachys. Schuttlw.		Dry sandy hills.
" "	552	Chapmania Floridana. T. & G.	Leguminosae.	East Florida.
" "	553	Rhynchosia tomentosia. Torr. & Gray.		Dry sandy soil.

MONTH.	No.	NAME OF PLANT	NATURAL ORDER	LOCALITY
May to August	551	Rhynchosia tomentosa, var. monophylla. Torr. & Gray.	Leguminosae.	Dry sandy soil.
	555	Rhynchosia tomentosa. var. erecta. Torr. & Gray.		Dry sandy soil.
	556	Melothria pendula. Lin.	Cucurbitaceae.	Light soil.
	557	Diplopappus obovatus. Torr. & Gray.	Compositae.	Low pine barrens.
..	558	Berlandiera subacaulus. Nutt.	..	East Florida.
..	559	Cacalia diversifolia. T. & G	..	Muddy banks of Chipola river.
..	560	Lobelia paludosa. Nutt.	Lobeliaceae.	Pine barren swamps.
..	561	Specularia perfoliata. D. C.	Campanulaceae.	Fields and low places.
..	562	Plantago major. Lin.	Plantaginaceae.	Around dwellings. Introduced.
..	563	'' Virginica. Lin.		Low sandy soil. Common.
..	564	Priva echinata. Juss.	Verbenaceae.	South Florida.
..	565	Verbena Aubletia. Lin.	..	Dry, light soil.
..	566	Scutellaria integrifolia. Lin.	Labiatae.	Dry sandy soil.
..	567	'' var. major.	..	Swamps.
..	568	Forsteronia difformis. A. D. C.	Apocynaceae.	River banks.
..	569	Echites umbellata. Jacqu.	..	South Florida.
..	570	'' Andrewsii.	..	Dry shores. South Florida.
..	571	Rumex crispus. Lin.	Polygonaceae.	Waste places around dwellings.
..	572	Saururus cernuus. Lin.	Saururaceae.	Marshy banks and places. Pal.
..	573	Stillingia ligustrina. Michx.	Euphorbiaceae.	River banks. Palatka.
..	574	Excoecaria lucida. Swartz.	..	South Florida.
	575	Tragia urens. Lin.	..	Dry sandy soil.
	576	Zannichellia palustris. Lin.	Naiadaceae.	Fresh or brackish water.
	577	Ruppia maritima. Lin.	..	Shallow waters along the coast.
	578	Eriocaulon gnaphaloides. Michx.	Eriocaulonaceae.	Boggy places. Pal to Br. L'ding.
	579	Rhynchospora megalocarpa. Gray.	Cyperaceae.	Dry sandy soil along E & W Fla.
	580	Chaetospora nigricans. Kunth.		Damp soil. Mariana, W. Fla.
	581	Scleria pauciflora. Muhl. var. glabra.		Low and sandy.
	582	Eustachys petraea. Desv.	Gramineae.	Damp soil along the coast.
May to Sept.	583	Callirhoë Papaver. Gray.	Malvaceae.	Rich open woods. Tallahassee.
.. ..	584	Malvastrum tricuspidatum. Gray.		South Florida. Indian river.
.. ..	585	Galium hispidulum. Michx.	Rubiaceae.	Dry sandy soil near the coast.
.. ..	586	Ilysanthes gratioloides. Benth.	Scrophulariaceae.	Springs and rivulets. Common.
.. ..	587	Lippia nodiflora. Michx.	Verbenaceae.	Damp sandy soil around Palatka
.. ..	588	Batatas littoralis. Choix.	Convolvulaceae.	Drifting sand along the coast.
.. ..	589	Calystegia spithamaea. Pursh.		Dry soil.
.. ..	590	Amaranthus albus. Lin.	Amaranthaceae.	Cultivated ground.
.. ..	591	Stillingia aquatica. n. sp.	Euphorbiaceae.	Pine barren ponds. Palatka.
.. ..	592	'' sylvatica. Lin.		Light sandy soil.
.. ..	593	Cnidoscolus stimulosus. Gray.	..	Dry pine barrens. Palatka.
.. ..	594	Crinum Americanum. Lin.	Amaryllidaceae	River swamps. Palatka.
.. ..	595	Juncus effusus. Lin.	Juncaceae.	Bogs and swamps. Palatka.
.. ..	596	Commelyna Virginica. Lin.	Commelynaceae.	Light sandy soil. Palatka.
.. ..	597	Fuirena scirpoidea. Vahl.	Cyperaceae.	Wet sandy places near the coast.
.. ..	598	Eleocharis tuberculosa. R. Br.	..	Wet places, chiefly along coast.
.. ..	599	'' simplex. Torr.		Miry places along streams.
.. ..	600	'' prolifera. Torr.		Marshy banks; ponds of streams
.. ..	601	'' albida. Torr.		Wet places along the coast.
.. ..	602	'' arenicola. Torr.		Sandy sea shores. West Fla.

MONTH.	No.	NAME OF PLANT.	NATURAL ORDER.	LOCALITY.
May to Sept.	603	Eleocharis tricostata. Torr.	Cyperaceae.	Low pine barrens.
" "	604	" microcarpa. Torr.	"	Low sandy places.
" "	605	" liliculmis. Torr.	"	Low sandy places.
" "	606	Sporobolus Indicus. Brown.	Gramineae.	Waste places. Common.
" "	607	Polygala Rugelii. Shuttlw.	Polygalaceae.	Flat pine barrens.
" "	608	" Reynoldii. n. sp.	"	St. Augustine. East Florida.
May to Octob'r	609	Hypericum myrtifolium. Lam.	Hypericaceae.	Pine barren ponds. Palatka.
" "	610	Erigeron Philadelphicum. Lin.	Compositae.	Low grounds.
" "	611	Dichondra repens. Forst.	Convolvulaceae.	Low grounds.
" "	612	" rep. var. Caroliniensis. Choix.	"	Low grounds.
" "	613	Euphorbia cyathophora. Jac.	Euphorbiaceae.	Around dwellings. South Fla.
" "	614	" trichotoma. H. B. K.	"	South Florida.
" "	615	" inaequilatera. Sond.	"	South Florida.
" "	616	Panicum sanguinale. Lin.	Gramineae.	Cultivated ground; waste places
" "	617	Arenaria diffusa. Ell.	Caryophyllaceae.	Shady banks. Palatka.
" "	618	Sagittaria simplex. Pursh.	Alismaceae.	Shallow ponds in pine barrens.
May to Nov.	619	Cyperus trachynotus. Torr.	Cyperaceae.	Dry sandy soil.
" "	620	Portulaca pilosa. Lin.	Portulacaceae.	Common around Palatka.
" "	621	" oleracea. Lin.	"	Common around Palatka.
" "	622	Trianthema monogyna. Lin.	"	Keys of South Florida.
" "	623	Cypselea humifusa. Turp.	"	South Florida.
" "	624	Sesuvium pentandrum. Ell.	"	Muddy saline coves along coast.
May to Dec.	625	" portulacastrum. Lin	"	Sandy or muddy places on coast.
June to July.	626	Agrostemma Githago. Lin.	Caryophyllaceae.	Grain fields. Introduced.
" "	627	Hibiscus incanus. Wendl.	Malvaceae.	Ponds and marshes.
" "	628	Tilia pubescens. Ait..	Tiliaceae.	Rich soil.
" "	629	Corchorus siliquosus. Lin.	"	Key West.
" "	630	Zanthoxylum Carolinianum. Lam.	Rutaceae.	Dry soil near the coast.
	631	Zanthoxylum Floridanum. Nutt.		South Florida.
	632	Zanthoxylum Pterota. H. B. & K.	"	South Florida.
	633	Vitis bipinnata. Torr. & Gray.	Vitaceae	Margins of swamps.
	634	" acida. Lin.		Key West.
	635	" incisa. Nutt.		Sandy shores. St. Vincent Island
	636	" indivisa. Willd.		Banks of rivers.
	637	" Caribaea. D. C.		South Florida.
	638	" aestivalis. Michx.		Rich woods. Palatka.
	639	" vulpina. Lin.		River banks. Palatka.
	640	Ampelopsis quinquefolia. Michx.		Low grounds. Palatka.
"	641	Sentia ferrea. Brongn.	Rhamnaceae.	South Florida.
"	642	Berchemia volubilis. D. C.	"	Swamps around Palatka.
"	643	Frangula Caroliniana. Gray.	"	Banks of rivers. Palatka.
"	644	Crotalaria sagittalis. Lin.	Leguminosae.	Barren sandy soil. Palatka.
"	645	Trifolium procumbens. Lin.	"	Cultivated ground. Introduced.
"	646	Amorpha herbacea. Walt.		Low pine barrens.
"	647	Thephrosia Virginiana. Pers.	"	Dry pine barrens. Palatka.
"	648	" spicata. Torr. & G.		Dry soil.
"	649	" hispidula. Pursh.		Dry sandy soil.
"	650	" chrysophylla. Pursh.		Dry pine barrens.
"	651	" ambigua. Curtis.		Dry sandy soil.
"	652	" angustissima. Shuttlw.		South Florida.
	653	Rhynchosia parvifolia. D. C.		South Florida.

MONTH.	No.	NAME OF PLANT.	NATURAL ORDER.	LOCALITY.
June to July.	654	Rhynchosia menispermoidea. D. C.	Leguminosae.	Charlotte Harbor, South Fla.
	655	Gleditschia triacanthos. Lin.	"	Rich woods.
	656	" monosperma. Walt.	"	Deep river swamps.
	657	Neptunia lutea. Benth.	"	Damp soil near coast. Key West.
	658	Rosa setigera. Michx.	Rosaceae.	Borders of swamps. Palatka.
	659	" Carolina. Lin.	"	Swamps.
	660	" laevigata. Michx.	"	Common.
	661	Laguncularia glabriflora. Presl.	Combretaceae.	Banks of Manatee river.
	662	Oenothera riparia. Nutt.	Onograceae.	Swamps; river banks. Palatka.
	663	Opuntia vulgaris. Mill.	Cactaceae.	Dry sandy soil.
	664	Mentzelia Floridana. Nutt.	Loasaceae.	South Florida.
	665	Piriqueta fulva. Chapm.	Turneraceae.	Dry light soil.
	666	" tomentosa. H. B. & K.	"	South Florida.
	667	" glabra. Chapm.	"	South Florida.
	668	Passiflora incarnata. Lin.	Passifloraceae.	In open ground, common. Pal.
	669	" lutea. Lin.	"	Woods and thickets. Palatka.
	670	" suberosa. Lin.	"	South Florida.
	671	" angustifolia. Swartz	"	South Florida.
	672	" Warei. Nutt.	"	South Florida.
	673	" multiflora. Lin.	"	Umbrella Key, South Florida
	674	Carica Papaya. Lin.	"	Indian River. South Florida.
	675	Hydrangea arborescens.	Saxifragaceae.	Banks of streams.
	676	Hydrocotyle interrupta. Muhl.	Umbelliferae.	Wet places. Palatka.
	677	Eryngium yuccaefolium. Michx.	"	Pine barrens, mostly damp soil.
	678	Daucus pusillus. Michx.	"	Dry sterile soil.
	679	Discopleura capillacea. D. C.	"	Brackish water.
	680	" Nuttalli. D. C.	"	Tampa bay.
	681	Thaspium trifoliatum. Gray.	"	Rich soil. Palatka.
	682	Galium uniflorum. Michx.	Rubiaceae.	Dry, rich soil. Palatka.
	683	" trifidum. Lin.	"	Wet places. Palatka.
	684	Spermacoce glabra. Michx.	"	Banks of rivers.
	685	Oldenlandia angustifolia. Gray	"	Sandy pine barrens.
	686	" ang. var. filifolia.	"	South Florida.
	687	Spigelia loganioides. A. D. C.	"	Near Fort King. East Florida.
	688	Valeriana scandens. Lin.	Valerianeae.	East Florida.
	689	Vernonia ovalifolia. Torr. & Gray.	Compositae.	Dry rich woods. Middle Florida.
	690	Pluchea pycnostachyum. Ell.	"	Damp pine barrens.
	691	Helianthella grandiflora. Torr. & Gray.		East Florida.
	692	Helianthella tenuifolia. Torr. & Gray.		Dry sandy pine barrens. W. Fla.
	693	Actinomeris pauciflora. Nutt.	"	Low pine barrens near coast.
	694	Cirsium repandum. Michx.	"	Dry pine barrens.
	695	Bejaria racemosa. Vent.	Ericaceae.	Dry sandy soil around Palatka.
	696	Bumelia lycioides. Gärtn.	Sapotaceae.	River banks.
	697	" lanuginosa. Pers.	"	Dry sandy soil.
	698	" reclinata. Vent.	"	River banks.
	699	Sideroxylon pallidum. Sprengl	"	South Florida.
	700	Dipholis salicifolia. A. D. C.	"	Indian River.
	701	Mimusops Sieberi. A. D. C.	"	South Florida.
	702	Pentstemon pubescens. Solander.	Scrophulariaceae.	Dry open woods around Palatka.
	703	Gratiola officinalis. Lin.		Swamps. Palatka suburbs.

MONTH.	No.	NAME OF PLANT	NATURAL ORDER.	LOCALITY.
June to July.	301	Dipteracanthus riparius. n. sp.	Acanthaceae.	Shady banks. Middle Florida.
" "	302	Callicarpa Americana. Lin.	Verbenaceae.	Dry open woods.
" "	303	Avicennia oblongifolia. Nutt.	"	Key West.
" "	304	" tomentosa. Jacq.	"	South Florida.
" "	305	Phlox maculata. Lin.	Polemoniaceae.	Low woods. Palatka.
" "	306	" nitida. Ell.	"	Low woods. Palatka.
" "	307	Cuscuta arvensis. Beyrich.	Convolvulaceae.	Fields; sterile soil on fine herbs
" "	308	Limnanthemum lacunosum. Griesb.	Gentianaceae.	Shallow ponds near Palatka.
" "	309	Apocynum androsaemifolium. Lin.	Apocynaceae.	Rich soil. Middle and West Fla.
" "	310	Asclepias tomentosa. Ell.	Asclepiadaceae.	Dry sandy pine barrens.
" "	311	" viridula. n. sp.	"	Pine barren swamps.
" "	312	" paupercula. Michx.	"	Marshes.
" "	313	" obtusifolia. Michx.	"	Sandy soil.
" "	314	" tuberosa. Lin.	"	Light dry soil. Palatka.
" "	315	Acerates viridiflora. Ell.	"	Dry sterile soil.
" "	316	" connivens. Decais.	"	Wet pine barrens.
" "	317	Rumex crispus. Lin.	Polygonaceae.	Waste grounds about dwellings.
" "	318	" Floridanus. Meissner.	"	Deep river swamps, W and S Fla.
" "	319	" hastatulus. Baldw.	"	Dry sand along coast. Mid. Fla.
" "	320	Euphorbia nudicaulis. n. sp.	Euphorbiaceae.	Low pine barrens near St. Joseph
" "	321	Sabal Palmetto. R. & S.	Palmae	Dry sandy soil around Palatka.
" "	322	" serrulata. R. & S.	"	Sandy soil around Palatka.
" "	323	" Adansonii. Guer.	"	Low grounds in lower districts.
" "	324	" Adans. var. megacarpa.	"	Dry rocky pine woods.
" "	325	Chamaerops Hystrix. Fraser.	"	Low shady woods in lower dists.
" "	326	Thrinax parviflora. Swartz.	"	Miami. South Florida.
" "	327	" argentea. Loddiger.	"	Coast and Keys, South Florida.
" "	328	" Garberi. Chapm.	"	Rocky pine woods. Miami S. Fla.
" "	329	Cocos nucifera. Lin.	"	Indian river. South Florida.
" "	330	Oreodoxa regia. H. B. K.	"	Rogers riv. east of Caximbar bay
" "	331	Liparis liliifolia. Rich.	Orchidaceae.	Low shady woods. Palatka sub's
" "	332	Calopogon pulchellus. R. Brown.	"	Swamps. Palatka's vicinity.
" "	333	Tillandsia utriculata. Leconte.	Bromeliaceae.	Ocklawaha river.
" "	334	" bracteata. n. sp.	"	South Florida.
" "	335	Iris tripetala. Walt.	Iridaceae.	Pine barren swamps.
" "	336	Allium Canadense. Kalm.	Liliaceae.	Banks of rivers.
" "	337	Zygadenus glaberrimus. Michx.	Melanthaceae.	Pine barren swamps.
" "	338	Stenanthium angustifolium. Gray.		Shady woods and banks.
" "	339	Mayaca Michauxii. Schott. & Endl.	Mayacaceae.	Springy places.
" "	340	Rhynchospora plumosa.	Cyperaceae.	Low pine barrens.
" "	341	" plum. var. intermedia.	"	Sandy barrens; often dry places.
" "	342	Rhynchospora oligantha. Gray.		Low, open pine woods.
" "	343	Rhynchospora rariflora. Ell.	"	Low grassy pine barrens.
" "	344	" compressa. Carey.	"	Margins pineb'n ponds. W. Fla.
" "	345	" stenophylla. n. sp.	"	Low grassy br'ns. Apalachicola.
" "	346	" decurrens. n. sp.	"	Marshy banks lakes and rivers.
" "	347	" patula. Gray.	"	Banks of pine barrens.
" "	348	" Elliottii. Dietr.	"	Margins ponds in pine barrens.

MONTH.	No.	NAME OF PLANT.	NATURAL ORDER.	LOCALITY.
June to July.	552	Rhynchospora miliacea. Gray.	Cyperaceae.	Bogs and deep miry places.
" "	553	" Grayii. Kunth.	"	Dry pine barrens.
" "	554	" Baldwinii. Gray.	"	Wet pine barrens.
" "	555	" ciliata. Vahl.	"	Wet pine barrens.
" "	556	" fascicularis. Nutt.	"	Low pine barrens.
" "	557	" pusilla. n. sp.	"	Margins barren ponds. Mid Fla.
" "	558	" divergens. n. sp.	"	Low pine barrens.
" "	559	Scleria oligantha. Ell.	"	Thickets and margins of fields.
" "	560	" Baldwinii. Torr.	"	Pine barren swamps.
" "	561	" gracilis. Ell.	"	Low pine barrens.
" "	562	" verticillata. Muhl.	"	Damp soil.
" "	563	Agrostis scabra. Willd.	Gramineae.	Sterile soil.
" "	564	" alba. Lin.	"	Damp soil.
" "	565	Bromus ciliatus Lin.. var purgans. Gray.	"	River banks and rich soil.
" "	566	Danthonia spicata. Beauv.	"	Dry barren soil.
" "	567	Aristida stricta. Michx.	"	Dry sandy ridges in pine barrens
June to Aug'st.	568	Thalictrum Cornuti. Lin.	Ranunculaceae.	Meadows, woods about Palatka.
" "	569	Cocculus Carolinus. D. C.	"	Woods and thickets. Common.
" "	570	Cabomba Carolineana. Gray.	Cabombaceae.	Ponds and still water.
" "	571	Hypericum maculatum. Walt.	Hypericaceae.	Dry pine barrens.
" "	572	" angulosum. Michx.	"	Pine barren ponds.
" "	573	" mutilum. Lin.	"	Ditches, low grounds. Common.
" "	574	" Sarothra. Michx.	"	Sandy old fields.
" "	575	Polygala lutea. Lin.	Polygalaceae.	Low pine barrens. Palatka.
" "	576	" Chapmanii. Torr. & Gray.	"	Low pine barrens near the coast.
"	577	Polygala incarnata. Lin.		Dry, sandy soil. Palatka.
"	578	Melilotus officinalis. Willd.	Leguminosae.	Waste places, fields. Introduced
	579	" alba. Lam.		Waste places, fields. Introduced
"	580	Zornia tetraphylla. Michx.		Sandy places along the coast.
"	581	Stylosanthes elatior. Swartz.		Sandy pine barrens.
"	582	Galactia Floridana. Torr. & Gray.		Dry sandy pine barrens.
	583	Galactia sessiliflora. Torr. & Gray.		Dry pine barrens.
	584	Schrankia uncinata. Willd.		Dry sandy soil.
	585	" angustata var. brachycarpa. Chapm.		Dry pine barrens.
"	586	Rhexia glabella. Michx.	Melastomaceae.	Low pine barrens.
"	587	Hypobrychia Nuttalli. Torr. & Gray.	Lythraceae.	Ponds and still waters. W. Fla.
	588	Laguncularia racemosa.	Combretaceae.	South Florida.
	589	Gaura angustifolia. Michx.	Onagraceae.	Dry old fields and sandy places near coast.
"	590	Proserpinaca palustris. Lin.	"	Ponds and ditches about Palatka
"	591	" pectinacea. Lam.	"	Ponds and ditches about Palatka
"	592	Sicyos angulatus. Lin.	Cucurbitaceae.	River banks. Palatka.
"	593	Borreria micrantha. Torr. & Gray.	Rubiaceae.	Waste places around Palatka.
	594	Vernonia angustifolia. Michx.	Compositae.	Dry sandy soil.
	595	Elephantopus Carolinianus.	"	Dry sandy soil. Palatka.
	596	Aster spinulosus. n. sp.	"	Damp pine barrens.
	597	" eryngiifolius. Torr. & Gray.	"	Low pine barrens.

MONTH.	No.	NAME OF PLANT.	NATURAL ORDER.	LOCALITY.
June to Aug'st.	798	Berlandiera tomentosa. Torr. & Gray.	Compositae.	Dry pine barrens.
	799	Echinacea atrorubens. Nutt.	"	Low pine barrens.
	800	Sonchus asper. Vill.	"	Fields. Common.
	801	Gratiola quadridentata. Michx	Scrophulariaceae.	Margins pine br'n ponds Palatka
	802	" pilosa. Michx.	"	Low grounds around Palatka.
	803	Dipteracanthus oblongifolius.	Acanthaceae.	Dry sandy pine barrens.
	804	Ruellia humistrata. Michx.	"	Grassy places around Palatka.
	805	Nepeta Cataria. Lin.	Labiatae.	Waste grounds. Introduced.
	806	" Glechoma. Benth.	"	Low shady places near dwel'ngs
	807	Physostegia Virginiana. Benth.	"	Low grounds and swamps.
	808	Leonitis nepetaefolia. R. Brown	"	Waste grounds around Palatka
	809	Heliotropium Curassavicum. Lin.	Borraginaceae.	Saline marshes. South Florida
	810	Heliotropium myosotoides. n. s		Palatka to South Florida.
	811	Capsicum frutescens. Lin.	Solanaceae.	South Florida.
	812	Asclepias cinerea. Walt.	Asclepiadaceae.	Flat sandy pine barrens Palatka
	813	Aristolochia Serpentaria. Lin.	Aristolochiaceae.	Shady woods.
	814	Euphorbia Floridana. n. sp.	Euphorbiaceae.	Dry pine barrens. Middle Fla.
	815	Parietaria debilis. Forst.	Urticaceae.	Damp shaded soil near the coast.
	816	Potamogeton perfoliatus. Lin.	Naiadaceae.	Fresh and brackish water.
	817	" fluitans. Roth.	"	Fresh water ponds and streams.
	818	Canna flaccida. Roscoe.	Cannaceae.	Ponds. Palatka, Leesburg.Apal.
	819	Juncus acuminatus. Ell.	Juncaceae.	Bogs and ditches about Palatka.
	820	Eleocharis multiflora. n. sp.	Cyperaceae.	Margins ponds and streams.
	821	Eriophorum Virginianum. Lin.	"	Bogs. swamps Palatka to Rice c.
	822	Ceratoschoenus capitatus. n. s.	"	Pine barren ponds. Middle Fla.
	823	Scleria triglomerata. Michx.	"	Low grounds around Palatka.
	824	" ciliata. Michx.	"	Dry pine barrens.
	825	Setaria composita. Kunth.	Gramineae.	Dry sandy soil coast South Fla.
June to Sept.	826	Ascyrum Crux Andreae. Lin.	Hypericaceae.	Sterile soil.
"	827	Phaseolus diversifolius. Pers.	Leguminosae.	Sandy soil around Palatka.
"	828	" helvolus. Lin.	"	Woods and margins of fields.
"	829	Centrosema Virginiana.	"	Light sandy soil around Palatka
"	830	Jussiea leptocarpa. Nutt.	Onagraceae.	Marshes.
"	831	Galium pilosum. Ait.	Rubiaceae.	Dry soil.
"	832	Diodia Virginiana. Lin.	"	Wet places.
"	833	Mitreola petiolata. Torr. & G.	"	Muddy banks.
"	834	Polypremum procumbens. Lin.	"	Waste places.
"	835	Coreopsis auriculata. Lin.	Compositae.	Rich shaded soil. Palatka.
"	836	Kalmia hirsuta. Walt.	Ericaceae.	Flat pine barrens about Palatka.
"	837	Herpestis Monnieria. Kunth.	Scrophulariaceae.	Ditches muddy banks near coast
"	838	Ruellia strepens. Lin.	Acanthaceae.	Dry rich soil around Palatka.
"	839	Dicliptera Halei. Riddell.	"	Shady banks of rivers.
"	840	" assurgens. Juss.		South Florida.
"	841	Evolulus diffusus. n. sp.	Convolvulaceae.	South Florida.
"	842	Solanum Carolinense. Lin.	Solanaceae.	Dry. waste places. Palatka.
"	843	" aculeatissimum. Jacqu.	"	Waste places. Palatka.
"	844	Polygonum dumetorum. Lin.	Polygonaceae.	Low margins fields and thickets.
"	845	Tragia urticifolia. Michx.	Euphorbiaceae.	Dry soil. Palatka's vicinity.
"	846	Croton balsamiferum. Willd.	"	South Florida.
"	847	" argyranthemum. Michx		Dry sandy pine barrens. Palatka
"	848	" monanthogynum. Michx		Dry sterile soil.
"	849	Aphora Blodgettii. Torr.		South Florida.
"	850	Drypetes crocea. Poir.		South Florida.
"	851	" glauca. Vahl.		South Florida.

MONTH.	No	NAME OF PLANT.	NATURAL ORDER.	LOCALITY.
June to Sept.	852	Batis maritima. Lin.	Batidaceae.	Salt marshes. Apalachicola.
" "	853	Sagittaria falcata. Pursh.	Alismaceae.	Lakes and rivers. Palatka.
" "	854	" natans. Michx.	"	Shallow ponds and streams.
" "	855	Tillandsia usneoides. Lin.	Bromeliaceae.	Humid situations. Palatka.
" "	856	Eleocharis melanocarpa. Torr.	Cyperaceae.	Pine barren swamps.
" "	857	" capitata. R. Brown.	"	Springy or miry places.
" "	858	" palustris. R. Brown.	"	Marshes and wet places.
" "	859	" obtusa. Schult.	"	Muddy marshes of ponds strems.
" "	860	" acicularis. R. Brown	"	Margins of ponds.
" "	861	" Baldwinii. Torr.	"	Swamps in Palatka's vicinity.
" "	862	Scirpus pungens. Vahl.	"	Sandy marshes along the coast.
" "	863	" Olneyi. Gray.	"	Brackish marshes W. Florida.
" "	864	Isolepsis capillaris. R. Brown.	"	Moist sandy places. Palatka.
" "	865	Eragrostis ciliaris. Link.	Gramineae.	Waste places near roads Key W.
" "	866	" Purshii. Schrad.	"	Waste places; cult. soil. Palatka
" "	867	Paspalum ciliatifolium. Michx	"	Wet and dry soil. Palatka.
" "	868	Stenotaphrum Americanum. Schrank.	"	Damp sandy places along coast.
June to Oct.	869	Siphonychia Americana. Torr. & Gray.	Caryophyllaceae.	Sandy banks of rivers.
" "	870	Siphonychia diffusa. n. sp.	"	Dry sandy pine barrens.
" "	871	Borrichia frutescens. D. C.	Compositae.	Saline marshes.
" "	872	Micranthemum orbiculatum. Michx.	Scrophulariaceae.	Muddy banks.
" "	873	Heliophytum Indicum. D. C.	Borraginaceae.	Waste places. South Florida.
" "	874	" parviflorum. D. C.	"	Waste places. South Florida.
" "	875	Evolvulus sericens. Schwartz.	Convolvulaceae.	Damp soil.
" "	876	" glabriusculus. Choix	"	South Florida.
" "	877	Podostigma pubescens. Ell.	Asclepiadaceae.	Low pine barrens.
" "	878	Metastelma Schlechtendalii. Decais.	"	South Florida.
" "	879	Metastelma parviflorum. R. Brown.	"	South Florida.
" "	880	Sarcostemma crassifolium. Decais.	"	South Florida.
" "	881	Seutera maritima. Decais.	"	Salt marshes.
" "	882	Alternanthera achyrantha. R. Brown.	Amarantaceae.	Along roads much trodden.
" "	883	Telanthera maritima. Moquin.	"	South Florida.
" "	884	Polygonum sagittatum. Lin.	Polygonaceae.	Wet places. Palatka.
" "	885	Euphorbia maculata.	Euphorbiaceae.	Cultiv'd grounds. Very common
" "	886	Ricinus communis. Lin.	"	Waste places. Common.
" "	887	Hippomane Mancinella. Lin.	"	South Florida.
June to Nov.	888	Hypericum Canadense. Lin.	Hypericaceae.	Ditches, low grounds. Common.
" "	889	Siphonychia erecta. n. sp.	Caryophyllaceae.	Sandy places along the coast.
" "	890	Paronychia Rugelii. Shuttl.	"	East Florida.
" "	891	Sida stipulata. Cav.	Malvaceae.	Waste places around dwellings.
" "	892	Lantana involucrata var. Fla.	Verbenaceae.	South Florida.
" "	893	Citharexylum villosum. Jacq.	"	South Florida.
" "	894	Duranta Plumieri. Jacq.	"	South Florida.
July.	895	Clematis Catesbyana. Pursh.	Ranunculaceae.	Dry sandy soil near the coast.
"	896	Nelumbium luteum. Willd.	Nelumbiaceae.	Lakes and still waters. Palatka.
"	897	Brasenia peltata. Pursh.	Cabombaceae.	Ponds and slow flowing streams
"	898	Nasturtium lacustre. Gray.	Cruciferae.	Rivers and cool springs. W. Fla.
"	899	" officinale. R. Brown.	"	Ditches. Common. Palatka.
"	900	Hibiscus aculeatus. Walt.	Malvaceae.	Margins swamps and ponds.

MONTH.	No.	NAME OF PLANT.	NATURAL ORDER.	LOCALITY
July.	901	Hibiscus grandiflorus. Michx.	Malvaceae.	Marshes near the coast.
"	902	Linum Virginianum. Lin.	Linaceae.	Sterile soil.
"	903	Ceanothus Americanus. Kin.	Rhamnaceae.	Dry woods. Palatka.
"	904	Polygala cymosa. Walt.	Polygalaceae.	Pine barren ponds. Palatka.
"	905	Psoralea virgata. Nutt.	Leguminosae.	Palatka's vicinity.
"	906	Rhynchosia minima. D. C.	"	Damp soil along the coast.
"	907	Baptisia simplicifolia. Croom.	"	Dry pine barrens near Quincy.
"	908	" microphylla. Nutt.	"	West Florida.
"	909	Ludwigia arcuata. Walt.	Onagraceae.	Muddy margins of ponds.
"	910	Myriophyllum laxum. Shuttl.	"	Ponds and lakes. Middle Fla.
"	911	" verticillatum. Lin.	"	Still water. Palatka.
"	912	Myriophyllum heterophyllum. Michx.	"	Ponds and ditches. Palatka.
	913	Hydrocotyle repanda. Pers.	Umbelliferae.	Low grounds. Palatka.
	914	Crantzia lineata. Nutt.	"	Muddy banks near the coast.
	915	Erynchium Virginianum. Lam.	"	Marshes. Common.
	916	Cicuta maculata. Lin.	"	Marshes. Palatka's vicinity.
	917	Sium lineare. Michx.	"	Along streams near Palatka.
	918	Archangelica hirsuta. Torr. & Gray.	"	Dry hills.
	919	Galium circaezans. Michx.	Rubiaceae.	Dry open woods. Palatka.
	920	Spermacoce Chapmannii. Torr. & Gray.	"	Dry soil.
	921	Spermacoce tenuior. Lin.		Dry soil.
	922	Borreria podocephalus. D. C.		Pine Key. South Florida.
	923	Oldenlandia Boscii. D. C.	"	River banks. Palatka.
	924	" glomerata. Michx.	"	Wet places.
	925	" Halei. Torr. & Gray.	"	River banks.
	926	Sclerolepis verticillata. Cars.	Compositae.	Shallow ponds.
	927	Tetragonotheca helianthioides. Lin.	"	Dry sandy soil.
	928	Rudbeckia nitida. Nutt.	"	Borders of swamps, thickets.
	929	Clethra alnifolia. Lin.	Ericaceae.	Swamps. Palatka.
	930	Cyrilla racemiflora. Walt.	Cyrillaceae.	Shady banks pine barren ponds.
	931	Pentstemon Digitalis. Nutt.	Scrophulariaceae.	Dry soil.
	932	Dipteracanthus linearis. T. & G	Acanthaceae.	South Florida.
	933	Dianthera ovata. Walt.	"	Muddy banks of streams.
	934	Scutellaria Floridana. n. sp.	Labiatae.	Pine barren swamps near coast.
	935	Stachys Floridana. Schuttle.	"	Low grounds. Middle to S. Fla.
	936	Gilia coronopifolia. Pers.	Polemoniaceae.	Dry sandy soil. Palatka.
	937	Sabbatia lanceolata. Torr. & Gray.	Gentianaceae.	Wet pine barrens. Palatka.
	938	Asclepias paniculata. Decais.	Asclepiadaceae.	Dry pine barrens.
	939	" longifolia. Ell.	"	Low pine barrens.
	940	Polygonum Persicaria. Lin.	Polygonaceae.	Low places around dwellings.
	941	Persea Caroliensis. Nees.	Lauraceae.	Rich shady woods.
	942	Laurus Catesbyana. Michx.	"	South Florida.
	943	Acalypha corchorifolia. Willd.	Euphorbiaceae.	South Florida.
	944	Sparganium ramosum. Huebn.	Typhaceae.	Lagoons and ditches. Palatka.
	945	Potamogeton hybridus. Michx.	Naiadaceae.	Shallow ponds.
	946	Echinodorus rostratus. Engel.	Alismaceae.	South Florida.
	947	Bletia verecunda. Swartz.	Orchidaceae.	Open pine barrens. Mid. & E. Fla
	948	Gymnadenia nivea. Gray & Engel.	"	Pine barren swamps. Palatka.
"	949	Listera australis. Lindl.	"	Wet, shady woods. Palatka.
"	950	Agave Virginica. Lin.	Amaryllidaceae.	Sterile soil.

MONTH	No.	NAME OF PLANT.	NATURAL ORDER.	LOCALITY
July	951	Lophiola aurea. Lin.	Haemodoraceae	Wet pine barrens. Palatka to Brown's Landing.
	952	Dioscorea villosa. Lin.	Dioscoreaceae	Margins of swamps. Common.
	953	Cephaloxis flabellata. Decais.	Juncaceae.	Miry banks ponds and streams
	954	Xyris Elliottii. n. sp.	Xyridaceae.	Wet grassy pine barrens.
	955	" difformis. n. sp.	"	Low pine barrens near coast.
	956	Eleocharis elongata. n. sp.	Cyperaceae	In still water. Apalachicola
	957	Psilocarya rhynchosporoides. Torr.	"	Shallow pine barren ponds.
	958	Zizania aquatica. Lin.	Gramineae	Deep marshes and ponds.
	959	Brachyelytrum aristatum. Beauv.		Dry rocky places.
	960	Glyceria nervata. Trin.		Wet places; swamps. West Fla.
	961	" pallida. Trin.		Shallow water.
	962	Lolium arvense. Withering.		Apalachicola.
July & Aug'st.	963	Lechea major. Michx.	Cistaceae.	Dry sterile soil. Palatka.
"	964	" minor. Michx.	"	Dry sterile soil. Palatka.
" "	965	Hypericum fasciculatum. Lam	Hypericaceae.	Margins of pine barren ponds.
" "	966	" galioides. Lam.	"	Pine barrens.
" "	967	" galioides. var. ambiguum.	"	River swamps.
	968	Hypericum nudiflorum. Michx		Low grounds.
	969	" pilosum. Walt.		Wet pine barrens.
	970	" Drummondii. Torr. & Gray.		Dry pine barren soil.
	971	Elodea Virginica. Nutt.	"	Swamps.
	972	" petiolata. Pursh.	"	Swamps.
	973	Hibiscus coccineus. Walt.	Malvaceae.	Deep marshes. Palatka.
	974	" Floridanus. Schuttl.	"	South Florida. Indian river.
	975	" tiliaceus. Lin.	"	Indian river. South Florida.
	976	Gordonia Lasianthus. Lin.	Camelliaceae.	Swamps. Palatka.
	977	" pubescens. L'Her.	"	Near the coast.
	978	Rhus glabra. Lin.	Anacardiaceae.	Open woods dry rich soil. W. Fla
	979	" copallina. Lin.	"	Margins fields and open woods.
	980	" venenata. D. C.	"	Swamps.
	981	" Toxicodendron. Lin. var. quercifolium. Torr. & Gray	"	Dry pine barrens.
	982	Rhus Toxicodendron var. radicans. Torr.		Swamps.
	983	Rhus Metopium. Lin.		Swamps.
	984	" aromatica. Ait.	"	Dry, open woods. West Florida
	985	Polygala Baldwinii. Nutt.	Polygalaceae.	Low pine barrens.
	986	Indigofera Caroliniana. Walt.	Leguminosae	Dry pine barrens. Palatka.
	987	" leptosepala. Nutt.	"	South Florida.
	988	Desmodium acuminatum. D. C.	"	Rich, shady soil. Palatka.
	989	" nudiflorum. D. C.	"	Rich woods.
	990	" canescens. D. C.		Dry, open woods. Palatka.
	991	" cuspidatum. D. C.		Dry, open woods.
	992	" Floridanum. n. sp.		Dry sandy hills. Apalachicola.
	993	" tenuifolium. Torr. & Gray.		Dry pine barrens.
	994	Apios tuberosa. Moench.		Swamps. Palatka.
	995	Phaseolus perennis. Walt.		Low woods and margins fields
	996	" sinuatus. Nutt.		Dry sandy ridges pine barrens.
	997	Clitoria Mariana. Lin.		Light or sandy soil. Palatka.
	998	" Schreiberii. Schauffrk.		Rich woods. Palatka

PROBABLY ALMANAC.

MONTH.	N	NAME OF PLANT	NATURAL ORDER.	LOCALITY
July & Aug st		Cracca pilosa. Ell.	Leguminosae	Dry soil.
"		mollis. Michx.		Dry sandy pine barrens
	1001	glabella. Michx.		Dry sandy pine barrens
	1002	brachypoda. Torr. & Gray.		Dry sandy ridges in pine barrens
	1003	Cassia obtusifolia. Lin.		Waste places. Palatka.
	1004	Chamaecrista. Lin		Dry barren soil.
	1005	Mimosa strigillosa. Torr. &		Near Palatka. Common.
	1006	Rhexia stricta. Pursh.	Melastomaceae.	Margins ponds in pine barrens
	1007	ciliosa. Michx.	"	Bogs in the pine barrens.
	1008	serrulata. Nutt.	"	Open, flat pine barrens.
	1009	lutea. Walt.		Pine barren swamps. Palatka.
	1010	Ludwigia virgata. Michx.	Onagraceae	Low pine barrens.
	1011	capitata. Michx.		Wet pine barrens.
	1012	alata. Ell.	"	Brackish marshes.
	1013	microcarpa. Michx.	"	Muddy places.
	1014	palustris. Ell.	"	Ditches; muddy places. Common
	1015	spathulata. Torr & Gray.	"	Margins p. b. ponds. Middle Fla
	1016	Eryngium Cervantesii. Laroch.	Umbelliferae.	Damp sandy soil along W. coast.
	1017	Aralia spinosa. Lin.	Araliaceae.	Swamps. Palatka.
	1018	Cephalanthus occidentalis. Lin	Rubiaceae.	Ponds and marshes. Palatka.
	1019	Liatris squarrosa. Willd.	Compositae.	Dry, sandy soil. Palatka.
	1020	Polymnia Uvedalia. Lin.		Rich soil.
	1021	Rudeckia hirta. Lin.	Compositae.	Dry soil.
	1022	laciniata. Lin.	"	Swamps.
	1023	Marshallia angustifolia. Pursh.	"	Low pine barrens.
	1024	Cacalia ovata. Walt.	"	Swamps.
	1025	Cirsium Nuttallii. D. C.		Dry, light soil.
"	1026	Lecontei. Torr. & Gray.		Pine barren swamps.
"	1027	Rhododendron viscosum. Torr.	Ericaceae.	Swamps. Palatka.
"	1028	Mimulus alatus. Ait.	Scrophulariaceae.	Swamps. Palatka.
"	1029	Buchnera elongata. Swartz.		Low pine barrens.
"	1030	Americana. Lin.		Low pine barrens.
"	1031	Dipteracanthus noctiflorus. Nees.	Acanthaceae.	Low grassy pine barrens.
	1032	Dicliptera brachiata. Sprengl.	"	River banks.
	1033	Phryma leptostachya. Lin.	Verbenaceae.	Rich shaded soil. Palatka
	1034	Ocimum Campechianum. Mil.	Labiatae.	South Florida.
	1035	Micromeria Brownei. Benth.	"	River banks.
	1036	Salvia azurea. Lam.		Dry light or sandy soil. Palatka
	1037	Scutellaria canescens. Nutt. var. punctata.		Dry open woods. Palatka.
	1038	Scutellaria pilosa. Michx.		Dry sandy soil.
	1039	parvula. Michx.		Rocky woods. Southwest Fla.
	1040	Macbridea alba. n. sp.		Low pine barrens. West Fla.
	1041	Hydrolea quadrivalvis. Walt.	Hydroleaceae	Pools; muddy banks of streams
	1042	Nama Jamaicensis. Lin.	"	South Florida.
	1043	Sabbatia macrophylla. Hook.	Gentianaceae	Wet pine barrens.
	1044	brachiata. Ell.	"	Low grounds.
	1045	gracilis. Pursh	"	Low grassy pine b'ns. Palatka
	1046	calycosa. Pursh.		River swamps.
	1047	chloroides. Pursh.		Margins pine barren ponds.
	1048	gentianoides. Ell.		Low pine barrens. Palatka.

MONTH.	No.	NAME OF PLANT.	NATURAL ORDER.	LOCALITY.
July & Aug'st.	1049	Apocynum cannabinum. Lin. var. glaberrimum.	Apocynaceae.	Dry soil. West Florida.
	1050	Conolobus macrophyllus. Michx.	Asclepiadaceae.	Low thickets. Palatka.
	1051	Conolobus flavidulus. Chapn.	"	Light rich soil. Palatka.
	1052	" prostratus. Baldw.	"	Sand hills.
	1053	Phytolacca decandra. Lin.	Phytolaccaceae.	Margins fields; waste grounds.
	1054	Obione arenaria. Moquin.	Chenopodiaceae.	Drifting sands along the coast.
	1055	" cristata. Moquin.	"	Sandy shores. South Florida.
	1056	Chenopodium Boscianum. Moquin.	"	Cultivated grounds.
	1057	Chenopodium album. Lin.		Waste places.
	1058	" murale. Lin.		Waste grounds.
	1059	" Anthelminticum. Lin.		Waste grounds.
	1060	Salicornia ambigua. Michx.	"	Sandy marshes along the coast.
	1061	Chenopodina maritima. Moqui.	"	Low sandy places along coast.
	1062	Typha latifolia. Lin.	Typhaceae.	Margins ponds, rivers. Palatka.
	1063	Naias flexilis. Rostk.	Naiadaceae.	In ponds, still waters. Palatka.
	1064	" fusiformis. Rostk.	"	Brackish water along the coast.
	1065	Alisma Plantago. Lin.	Alismaceae.	Margins ponds, ditches. Palatka
	1066	Echinodorus parvulus. Engelm.	"	Margins shallow ponds. Mid. Fla
	1067	Microstylis ophioglossoides. Nutt.	Orchidaceae.	Low shady wood. Palatka.
	1068	Microstylis Floridana. n. sp.	"	Wet shady woods. Apalachicola
	1069	Bletia aphylla. Nutt.	"	Rich shaded soil. Palatka.
	1070	Pogonia pendula. Lindl.	"	Rich shaded woods. Palatka.
	1071	Gymnadenia flava. Lindl.	"	Open grassy swamp. Palatka.
	1072	Plantanthera flava. Gray.	"	Low shady banks.
	1073	Sisyrinchium Bermudiana. Lin	Iridaceae.	Grassy meadows; sometimes in dry soil. Palatka.
	1074	Smilax laurifolia. Lin.	Smilaceae.	Swamps; margins of ponds. Pal.
	1075	Melanthium Virginicum. Lin.	Melanthaceae.	Swamps.
	1076	Schollera graminea. Willd.	Pontederiaceae.	In flowing water. White Water.
	1077	Xyris elata. nova. sp.	Xyridaceae.	Sandy swamps near the coast.
	1078	" Caroliniana. Walt.	"	Shallow ponds and swamps.
	1079	Cyperus flavescens. Lin.	Cyperaceae.	Low ground. Palatka.
	1080	Rhynchospora semiplumosa. Gray.	"	Dry sandy ridges near the coast.
	1081	Rhynchospora microcarpa. Baldw.	"	Margins of ponds.
	1082	Rhynchospora filifolia. Gray.	"	Margins of pine barren grounds.
	1083	" gracilenta. Gray.	"	Wet pine barrens.
	1084	" cephalantha. Gray.	"	Bogs and shady swamps.
	1085	Cladium effusum. Torr.	"	Fresh or brackish coast marshes
	1086	Scleria interrupta. Michx.	"	Low pine barrens.
	1087	Leersia oryzoides. Swartz.	Gramineae.	Ditches and swamps.
	1088	" Virginica. Willd.	"	Swamps and margins of streams
	1089	" hexandra. Swartz.	"	In lakes and ponds. Palatka.
	1090	Hydrochloa Caroliniensis. Beauv.	"	Still water or creeping on muddy banks.
	1091	Sporobolus Virginicus. Kunth.	"	Saline marshes along coast.
	1092	Vilfa aspera. Beauv.	"	Dry sandy soil.
	1093	Agrostis perennis. Gray.	"	Swamps; river banks. Palatka.
	1094	Aristida lanata. Poir.	"	Dry pine barrens.
	1095	Spartina juncea. Willd.	"	Sandy, marshy places near coast

MONTH.	No.	NAME OF PLANT.	NAT. ORDER.	LOCALITY.
July & Aug'st.	1096	Spartina gracilis. Hook.	Gramineae.	Sandy saline swamps.
" "	1097	Ctenium Americanum. Sprengl	"	Low pine barrens.
" "	1098	Uniola latifolia. Michx.	"	Banks of rivers. Palatka.
" "	1099	" paniculata. Lin.	"	Drifting sand along coast.
" "	1100	" gracilis. Michx.	"	Rich damp soil.
" "	1101	Elymus Virginicus. Lin.	"	River banks. Palatka.
" "	1102	Paspalum Walteri. Schultes.	"	Low cultivated grounds.
" "	1103	" laeve. Michx.	"	Dry woods and margins fields.
July to Sept.	1104	Ascyrum stans. Michx.	Hypericaceae.	Damp soil. Palatka.
" "	1105	Sida spinosa. Lin.	Malvaceae.	Waste places. Palatka.
" "	1106	Kosteletzkya Virginica.	"	Marshes; low ground near coast.
" "	1107	Impatiens fulva. Nutt.	Balsaminaceae.	Shady swamps. Palatka.
" "	1108	Polygala ramosa. Ell.	Polygalaceae.	Low open pine barrens. Palatka
" "	1109	" Hookeri. Torr. & Gray.	"	Low grassy pine bar'ns. W. Fla.
" "	1110	" grandiflora. Walt.	"	Dry soil.
" "	1111	Desmodium strictum. D. C.	Leguminosae.	Pine barrens. Common.
" "	1112	Phaseolus Lauppii. Schaffrnk.	"	Rich shady woods. Palatka.
" "	1113	Vigna glabra.	"	Brackish marshes.
" "	1114	Canavalia obtusifolia. D. C.	"	Sandy shores St. Vincent Island.
" "	1115	Lythrum alatum. Pursh.	Lythraceae.	Swamps and river banks.
" "	1116	Gaura filipes. Spach.	Onagraceae.	Dry pine barrens.
" "	1117	Jussiaea decurrens. D. C.	"	Ditches around Palatka.
" "	1118	Ludwigia linearis. Walt.	"	Ditches and ponds.
" "	1119	" linifolia. Poir.	"	Swamps and ditches.
" "	1120	" cylindrica. Ell.	"	Swamps around Palatka.
" "	1121	" pilosa. Walt.	"	Ditches and ponds near coast.
" "	1122	" sphaerocarpa. Ell.	"	Margins of ponds.
" "	1123	" lanceolata. Ell.	"	Ponds and swamps. Palatka.
" "	1124	" natans. Ell.	"	Marshes and margins of streams
" "	1125	Diodia teres. Walt.	Rubiaceae.	Dry sandy soil.
" "	1126	Mitreola sessilifolia. Torr. & Gray.	"	Grassy swamps. Palatka to Brown's Landing.
"	1127	Vernonia Noveboracensis. Willd.	Compositae.	River banks and low grounds. Palatka.
"	1128	Liatris Chapmannii. Torr. & Gray.		Dry sandy ridges.
	1129	Eupatorium purpureum. Lin.		Swamps around Palatka.
	1130	Boltonia glastifolia. L'Her.		River swamps.
	1131	Silphium compositum. Michx.		Sandy open woods.
	1132	" Asteriscus. Lin.		Dry open woods.
	1133	Ambrosia artemisiaefolia. Lin.		Cultivated grounds. Common.
	1134	Zinnia multiflora. Lin.		Waste places.
	1135	Lepachys pinnata. Torr. & Gray.		Dry soil. West Florida.
	1136	Helianthus debilis. Nutt.		Shores of East Florida.
	1137	" praecox. Gray & Englm.		Sandy shores of West Florida.
	1138	Bidens frondosa. Lin.		Low grounds. Palatka.
	1139	Gaillardia lanceolata. Michx.		Dry pine barrens.
	1140	Palafoxia integrifolia. Torr. & Gray.		Dry pine barrens.
"	1141	Erechthites hieracifolia. Raf.		Rich soil. Common. Palatka.
"	1142	Lactuca elongata. Muhl. var. integrifolia.		Dry soil.
"	1143	Lactuca elongata. Muhl. var. graminifolia.		Dry soil.

MONTH	No	NAME OF PLANT	NATURAL ORDER	LOCALITY
July to Sept.	141	Lobelia cardinalis. Lin.	Lobeliaceae.	Muddy banks. Palatka.
" "	142	" Boykini. Torr & Gray.	"	Margins pine barrens ponds. Palatka.
	143	Utricularia cornuta. Michx.	Lentibulaceae.	Swamps. Palatka.
	144	Herpestis amplexicaulis. Pursh.	Scrophulariaceae.	Pine barrens ponds.
	145	Gratiola subulata. Baldw.		Low sandy pine barrens.
	146	Seymeria pectinata. Pursh.		Dry sandy soil.
	147	Darystoma quercifolia. Benth.		Rich woods and river banks.
	148	Verbena angustifolia. Michx.	Verbenaceae.	Dry woods.
	149	Hyptis radiata. Willd.	Labiatae.	Low grounds.
	150	Scutellaria lateriflora. Lin.	"	Shady swamps. Palatka.
	151	Teucrium Canadense. Lin.	"	Swamps; low grounds. Palatka
	152	Pharbitis Nil. Chois.	Convolvulaceae.	Cultivated grounds.
	153	Ipomoea perennis. Schaffranek	"	Rich shady soil. Palatka.
	154	" Michauxii. Sweet.	"	Light sandy soil.
	155	" sagittifolia. Bot. Reg.	"	Salt marshes.
	156	" fastigiata. Sweet.	"	South Florida.
	157	Jacquemontia violacea. Chois.	"	South Florida.
	158	Stylisma humistrata. Chapm.	"	Dry sandy pine barrens.
	159	" aquatica. Chapm.	"	Margins of ponds.
	160	" Pickeringii. Gray.	"	Sandy pine barrens.
	161	Solanum nigrum. Lin.	Solanaceae.	Waste places. Common.
	162	" Radula. Vahl.	"	South Florida.
	163	" mammosum. Lin.	"	Waste places. Palatka.
	164	Physalis angustifolia. Nutt.	"	Low sandy places. West Fla.
	165	Lycium Caroliniamum. Michx.	"	Salt marshes.
	166	Asclepias verticillata.	Asclepiadaceae.	Open woods. Palatka.
	167	Boerhaavia erecta. Lin.	Nyctaginaceae.	Cultivated grounds.
	168	" hirsuta. Willd.	"	South Florida.
	169	" viscosa. Lag.	"	South Florida.
	170	Pisonia aculeata. Lin.	"	South Florida.
	171	" obtusata. Swartz.	"	South Florida.
	172	Euxolus lividus. Moquin.	Amarantaceae.	South Florida.
	173	Alternanthera flavescens. Moquin.	"	Margins of fields. Middle Fla.
	174	Telanthera Floridana. n. sp.	"	South Florida.
	175	" Brasiliena. Moquin	"	South Florida.
	176	Oplotheca Floridana. Nutt.	"	Dry sandy places.
	177	Polygonum acre. Kunth.	Polygonaceae.	Ditches and margins of ponds.
	178	" hydropiperoides. Michx.	"	Ditches; muddy banks. Palatka
	179	Polygonum setaceum. Baldw.		Low ground. Palatka.
	180	" hirsutum. Walt.		Pine barren ponds.
	181	Eriogonum longifolium. Nutt.		Sand ridges.
	182	" tomentosum. Michx.		Dry pine barrens.
	183	Euphorbia corollata. Lin.	Euphorbiaceae.	Dry rich soil.
	184	" obtusata. Pursh.	"	Shady woods.
	185	" cordifolia. Ell.	"	Sandy pine barrens.
	186	Acalypha Virginica. Lin.	"	Fields and around dwellings.
	187	" gracilis. Gray.	"	Sterile soil.
	188	" Caroliniana. Walt.	"	Cultivated grounds.
	189	Croton ellipticum. Ell.	"	Pine barrens.
	190	" grandulosum. Lin.	"	Dry waste places.
	191	Pilea pumila. Gray.	Urticaceae.	Wet shaded places.
	192	Bochmeria cylindrica. Willd.	"	Swamp thickets. Palatka.

MONTH.	No.	NAME OF PLANT	NATURAL ORDER	LOCALITY
July to Sept.	1196	Zostera marina. Lin.	Naiadaceae	Deep salt water.
" "	1197	Echinodorus radicans. Englm.	Alismaceae	Swamps.
" "	1198	Sagittaria varibilis. Engelm.		Lakes and rivers. Lake Como
" "	1199	Vallisneria spiralis. Lin.	Hydrocharidaceae	Slow flowing streams. Palatka.
" "	1200	Limnobium Spongia. Richard.		Still water.
" "	1201	Pancratium maritimum. Lin.	Amaryllidaceae	Salt marshes.
" "	1202	Lachnanthes tinctoria. Ell.	Haemodoraceae	Ponds and ditches. Palatka to Rice Creek.
" "	1203	Juncus scirpoides. Lam.	Juncaceae	Sandy swamps.
" "	1204	" polycephalus. Ell.		Ponds; miry margins of streams
" "	1205	" abortivus. n. sp.		Grassy margins ponds near coast
" "	1206	" marginatus. Rostk.		Ditches and low grounds.
" "	1207	" biflorus. Ell.		Ditches and low grounds.
" "	1208	Pontederia cordata. Lin.	Pontederiaceae	Miry margins ponds and rivers.
" "	1209	" lancifolia. Muhl		Miry marg. ponds about Palatka
" "	1210	" angustifolia. Pursh.		Miry marg. ponds about Palatka
" "	1211	Xyris ambigua. Beyer.	Xyridaceae	Open grassy pine barrens.
" "	1212	" stricta. n. sp.		Shallow ponds in pine barrens.
" "	1213	" platylepis. n. sp.		Low sandy places.
" "	1214	" torta. Smith.		Sandy, often dry soil.
" "	1215	" tenuifolia. n. sp.		Open grassy pine bar'ns swamps.
" "	1216	Eriocaulon decangulare. Lin.	Eriocaulonaceae	Boggy places. Palatka.
" "	1217	Cyperus Nuttallii. Torr.	Cyperaceae	Salt or brackish soil.
" "	1218	" microdontus. Torr.		Margins of ponds and streams.
" "	1219	" strigosus. Lin.		Swamps and damp soil.
" "	1220	" stenolepis. Torr.		Swamps and wet places.
" "	1221	" repans. Ell.		Sandy soil near the coast.
" "	1222	" Lecontii. Torr.		Low sandy places along coast.
" "	1223	Cyperus virens. Michx.		Miry places. Common.
" "	1224	" inflexus. Muhl.		Low sandy places W. Florida.
" "	1225	" compressus. Lin.		Cultivated grounds. Common.
" "	1226	" filiculmis. Vahl.		Dry sandy soil.
" "	1227	" retrofractus. Torr.		Sandy barren soil.
" "	1228	" Baldwinii. Torr.		Cultivated grounds.
" "	1229	" erythrorhizos. Muhl.		Ponds and ditches.
" "	1230	Kyllingia pumila. Michx.		Wet places.
" "	1231	Lipocarpha maculata. Torr.		Springy or miry places.
" "	1232	Fuirena squarrosa. Michx.		Swamps. Palatka.
" "	1233	" hispida. Ell.		Swamps. Palatka.
" "	1234	Eleocharis equisetoides. Torr.		Ponds.
" "	1235	Scirpus lacustris. Lin.		Fresh or brackish marshes or ponds.
" "	1236	" Eriophorum. Michx.		Swamps; low grounds. Palatka
" "	1237	Rhynchospora glomerata. Vahl.		Bogs and springy places.
"	1238	Rhynchospora paniculata. Gray.		Bogs and springy places.
"	1239	Ceratoschoenus corniculatus. Nees.		Ponds and ditches.
" "	1240	Eustachys Floridana. n. sp.	Gramineae	Dry pine barrens. Middle Fla.
" "	1241	Paspalum Digitaria. Poir.		Open swamps.
" "	1242	Panicum gibbum. Ell.		Swamps.
" "	1243	" tenuiculmum. Meyer		South Florida.
" "	1244	" Walteri. Ell.		Damp soil.
" "	1245	" hians. Ell.		Old fields.
" "	1246	Setaria Italica. Kunth.		Swamps along the coast.

MONTH	No	NAME OF PLANT	NATURAL ORDER.	LOCALITY
July to Sept.	1246	Cenchrus echinatus. Lin.	Gramineae.	Fields and waste grounds.
"	1247	Tripsacum dactyloides. Lin.	"	Rich soil.
"	1248	" cylindricum. Michx.	"	Dry sandy soil.
July to Oct'b'r.	1250	Paronychia Baldwinii.	Caryophyllaceae.	Dry sandy soil. Middle Florida
"	1251	Sida rhombifolia. Lin.	Malvaceae.	Around dwellings. Common.
	1252	" ciliaris. Cav.	"	Key West.
	1253	" Elliottii. Torr. & Gray.	"	Open woods.
	1254	" Lindheimeri. Engelm. & Gray.		Key West.
	1255	Abutilon Hulseanum. Torr.		Tampa Bay.
	1256	" Jacquini. Don.		South Florida
	1257	" crispum. Gray.		Key West.
	1258	Modiola multifida. Mönch.		Waste places.
	1259	Polygala fastigiata. Nutt.	Polygalaceae.	Low pine barrens.
	1260	" cruciata. Lin.	"	Pine barren swamps.
	1261	" brevifolia. Nutt.	"	Bogs.
	1262	Krameria lanceolata. Torr.	Krameriaceae.	Tampa Bay.
	1263	Ipomoea tamnifolia. Lin.	Convolvulaceae.	Cultivated grounds. Common.
	1264	" sinuata. Ort.	"	Tampa. South Florida.
	1265	Cuscuta compacta. Juss.	"	Dry shady places.
	1266	Physalis viscosa. Lin.	Solanaceae.	Dry light soil.
"	1267	" lanceolata. Michx.	"	Dry sandy soil.
"	1268	" angulata. Lin.	"	Fields and waste grounds.
"	1269	Amarantus spinosus. Lin.	Amarantaceae.	Fields and waste places.
"	1270	" arborescens. Schaffrk.	"	Nova. spec. Shores St. Johns.
"	1271	Euphorbia polygonifolia. Lin.	Euphorbiaceae.	Drifting sand along the coast.
"	1272	Croton maritimum. Walt.	"	Drifting sand along the coast.
"	1273	Trichelostylis autumnalis.	Cyperaceae.	Low grounds.
"	1274	Cenchrus tribuloides. Lin.	Gramineae.	Sand along the coast.
Augu t.	1275	Petalostemon gracile. Kunth.	Leguminosae.	Low pine barrens.
"	1276	" carneum. Michx.	"	Dry sandy soil.
"	1277	Glottidium Floridanum. D. C.	"	Damp soil.
"	1278	Aeschynomene hispida. Willd.	"	Swamps.
"	1279	Lespedeza repens. Torr. & Gray.	"	Dry, sandy soil.
	1280	Lespedeza violacea. Pers.		Dry barren soil.
	1281	" divergens. Pursh.		Dry barren soil.
	1282	" sessiliflora. Michx.		Dry barren soil.
	1283	" reticulata. Pers.		Dry barren soil.
	1284	" hirta. Ell.		Dry barren soil.
	1285	" capitata. Michx.		Dry sterile soil.
	1286	Desmodium paucifolium. Nutt.		Shady woods. Palatka.
	1287	" viridiflorum. Beck.		Rich open woods.
	1288	" rotundifolium. D. C.		Dry open woods.
	1289	" Dillenii. Darl.		Open woods.
	1290	" laevigatum. D. C.		Dry rich soil.
	1291	" paniculatum. D. C.		Shady woods.
	1292	" Marilandicum. Boot		Dry open woods.
	1293	" rigidum. D. C.		Dry woods.
	1294	" lineatum. D. C.		Open grassy pine barrens.
	1295	Cassia Marilandica. L.		Rich soil.
	1296	Agrimonia Eupatoria. Lin.	Rosaceae.	Dry open woods.
	1297	" incisa. Torr. & Gray.	"	Dry open woods.
	1298	Lythrum lineare. Lin.	Lythraceae.	Brackish marshes. Palatka.
	1299	Nesaea verticillata. H. B. K.	"	Margins of ponds and marshes.
	1300	Cuphea aspera. n. spec.	"	Low pine barrens. St. Joseph.

MONTH.	No.	NAME OF PLANT.	NATURAL ORDER	LOCALITY
August.	1301	Cuphea microphylla. Schaflk.	Lythraceae.	Rich open woods. Palatka.
"	1302	Ludwigia alternifolia. Lin.	Onagraceae.	Sandy swamps.
"	1303	" hirtella. Raf.	"	Ditches and ponds.
"	1304	Erynchium virgatum. Lam.	Umbelliferae.	Pine barren swamps.
"	1305	" aromaticum. Baldw	"	Dry pine barrens. E. and S. Fla.
"	1306	Tiedemannia teretifolia. D. C.		Pine barren swamps.
"	1307	Liatris elegans. Willd.	Compositae.	Dry pine barrens.
"	1308	Brickellia cordifolia. Ell.	"	Light rich soil.
"	1309	Seriocoarpus tortifolius. Nees	"	Sandy pine barrens.
"	1310	Parthenium Hysterophorus. Lin.	"	Waste places. East Florida.
	1311	Rudbeckia heterophylla. Torr. & Gray.		Swamps. Middle Florida.
	1312	Rudbeckia maxima. Nutt.	"	Wet pine barrens.
	1313	Coreopsis semifolia.	"	Rich shaded woods.
	1314	Epiphegus Virginiana. Bart.	Orobranchaceae.	Under beach trees. Pal'ka H'gts
	1315	Elytraria virgata. Michx.	Acanthaceae.	Banks of rivers.
	1316	Verbena Caroliniana. Michx.	Verbenaceae.	Dry pine barrens.
	1317	" canescens. Kunth.	"	Apalachicola. West Florida.
	1318	Tournefortia gnaphaloides. R. Brown.	Borraginaceae.	Indian river. St. Lucy.
	1319	Tournefortia volubilis. Lin.	"	Indian river. South Florida.
	1320	Solanum Blodgettii.	Solanaceae.	South Florida.
	1321	Sabbatia paniculata. Pursh.	Gentianaceae.	Low grassy meadows. Palatka.
	1322	" angularis. Pursh.	"	Low rich grounds. Palatka.
	1323	Eustoma exaltatum. Grisb.	"	South Florida.
	1324	Celosia paniculata. Lin.	Amarantaceae.	South Florida.
	1325	Euphorbia Curtissii. Engl.	Euphorbiaceae.	Low pine barrens.
	1326	" commutata. Engl.	"	In suburbs of Apalachicola.
	1327	Potamogeton lucens. Lin.	Naiadaceae.	Fresh water. Apalachicola.
	1328	Tipularia discolor. Nutt.	Orchidaceae.	Shady banks.
	1329	Platanthera ciliaris. Lindl.	"	Swamps in the pine barrens.
	1330	" cristata. Lindl.	"	Bogs and swamps.
	1331	Habenaria Michauxii. Nutt.	"	Dry sandy soil.
	1332	Goodyera pubescens. R. Br.	"	Deep shady woods.
	1333	" repens. R. Br.	"	Shady woods.
	1334	" quercicola. Lindl.	"	Low shady woods.
	1335	Smilax lanceolata. Lin.	Smilaceae.	Rich woods; swamps. Palatka.
	1336	Cyperus rivularis. Kunth.	Cyperaceae.	Marshy banks streams. W. Fla.
	1337	" divergens. Kunth.	"	Cultivated streams. Middle Fla.
	1338	" Robbinsii. Oakes.	"	Shallow ponds. Quincy, Mid. Fla
	1339	" bicolor. n. sp.	"	Sandy margins near Quincy.
	1340	Scirpus divaricatus. Ell.	"	Muddy banks the Chipola river.
	1341	Abildgaardia monostachya. Vahl.	"	Indian river. South Fla.
	1342	Rhynchospora caduca. Ell.		Swamps in vicinity of Palatka.
	1343	Ceratoschoenus macrostachyus Gray.		Ponds and ditches.
"	1344	Leersia lenticularis. Michx.	Gramineae.	Ponds and swamps.
"	1345	Aristida purpurascens. Poir.	"	Dry pine barrens.
"	1346	" gracilis. Ell.	"	Dry soil.
"	1347	Trienspis ambigna.	"	Low pine barrens.
"	1348	Festuca nutans. Willd.	"	Rich woods and banks.
"	1349	Uniola nitida. Baldw.	"	Swamps.
Aug. & Sept.	1350	Hypericum cistifolium. Lam.	Hypericaceae.	Pine barren swamps near coast
" "	1351	Sesbania macrocarpa. Muhl.	Leguminosae.	Swamps.

MONTH.	No.	NAME OF PLANT.	NATURAL ORDER.	LOCALITY
Aug. & Sept.	1352	Aeschynomene viscidula. Michx.	Leguminosae.	Sandy places along the coast.
	1353	Amphicarpaea monoica. Nutt.		Rich soil in Palatka's vicinity.
	1354	Galactia spiciformis. Torr. & Gray.		South Florida.
	1355	Cassia biflora. Lin.	"	Key west. South Florida.
	1356	" aspera. Ell.	"	Dry old fields.
	1357	Ammannia humilis. Michx.	Lythraceae.	Ditches and muddy places.
	1358	" occidentalis. D. C. var. pygmaea.		Key West and South Florida.
"	1359	Archemora rigida. D. C.	Umbelliferae.	Swamps.
"	1360	Liatris spicata. Willd.	Compositae.	Swamps. Palatka suburbs.
"	1361	Mikania scandens. Willd.	"	Swamps in vicinity of Palatka.
"	1362	Iva frutescens. Nutt.	"	Saline marshes.
"	1363	" microcephala. Nutt.	"	Dry barren soil.
"	1364	" imbricata. Walt.	"	Drifting sands along the coast.
"	1365	Ambrosia trifida. Lin.	"	River banks; rich soil. Palatka.
"	1366	" erithmifolia. D. C.	"	Sandy shores. S. Fla., Key West
"	1367	Melanthera hastata. Michx.	"	Light, rich soil.
"	1368	" deltoida. Michx.	"	Indian river. South Florida.
"	1369	Heliopsis laevis. Pers.	"	Dry open woods.
"	1370	Rudbeckia fulgida. Ait.	"	Dry soil. Palatka.
"	1371	" triloba. Lin.	"	Dry soil.
"	1372	Actinomeris nudicaulis. Nutt.	"	Dry sandy woods.
"	1373	Coreopsis tripteris. Lin.	"	Woods and margins of fields.
"	1374	Bidens bipinnata. Lin.	"	Cultivated grounds. Common.
"	1375	Helenium autumnale. Lin.	"	Damp soil.
"	1376	Cacalia atriplicifolia. Lin.	"	Low grounds. West Florida.
"	1377	" tuberosa. Nutt.	"	Swamps.
"	1378	" lanceolata. Nutt.	"	Brackish marshes.
"	1379	Mulgedium Floridanum. D. C.	"	Rich soil.
"	1380	Lobelia puberula. Michx.	Lobeliaceae.	Swamps; low grounds. Palatka
"	1381	Campanula Americana. Lin.	Campanulaceae.	Dry rocky soil. W. to S. W. Fla.
"	1382	Monotropa uniflora. Lin.	Ericaceae.	Shady woods. Palatka.
"	1383	" Hypopitys. Lin.	"	Shady woods. Palatka.
"	1384	Statice Caroliniana. Walt.	Plumbaginaceae.	Salt marshes.
"	1385	Herpestis nigrescens. Benth.	Scrophulariaceae.	Low grounds.
"	1386	" peduncularis. Benth.	"	Key West. South Florida.
"	1387	Seymeria tenuifolia. Pursh.	"	Low pine barrens.
"	1388	Dasystoma pectinata. Benth.	"	Rich woods and river banks.
"	1389	Pedicularis Canadensis. Lin.	"	Shady woods and banks.
"	1390	Pycnanthemum incanum. Michx.	Labiatae.	Low ground.
	1391	Pycnanthemum aristatum. Michx.		Low ground.
	1392	Pycnanthemum muticum. Pers		Dry soil.
	1393	" linifolium. Pursh.		Dry soil.
	1394	Collinsonia anisata. Pursh.	"	Dry shaded soil.
	1395	Calamintha Caroliniana. Sweet		Sandy and rocky banks.
	1396	Trichostema dichotomum. Lin.	"	Dry sandy soil.
	1397	Calystegia sepium. R. Brown.	Convolvulaceae.	Rich soil.
	1398	Sabbatia Elliottii. Steud.	Gentianaceae.	Open pine barrens. Palatka.
	1399	" stellaris. Pursh.		Salt marshes.
	1400	Polygonella parvifolia. Michx.	Polygonaceae.	Dry sandy soil near the coast.
	1401	" ciliata. Meisner.	"	Manatee river. South Florida.

MONTH.	No	NAME OF PLANT.	NATURAL ORDER.	LOCALITY.
Aug. & Sept.	1102	Polygonum Virginianum. Lin.	Polygonaceae.	Dry rich soil. Palatka.
" "	1103	Euphorbia hypericifolia. Lin.	Euphorbiaceae.	St. Lucy. South Florida.
" "	1104	" glabella. Swartz.	"	Sandy sea shore. South Florida
" "	1105	" pilulifera. Lin.	"	Fort Myers. South Florida.
" "	1106	Crotonopsis linearis. Michx.	"	Dry sandy soil.
" "	1107	Phyllanthus Caroliniensis. Walt.	"	Low ground. South Florida.
	1108	Phyllanthus Niruri. Lin.	"	Fort Myers. South Florida.
	1109	Triglochin triandrum. Michx.	Alismaceae.	Salt marshes. West coast Fla.
	1410	Corallorhiza micrantha. n. sp.	Orchidaceae.	Shady woods.
	1411	Habenaria repens. Nutt.	"	Swamps: ditches. Mid. to S. Fla.
	1412	Lilium Catesbaei. Walt.	Liliaceae.	Low pine barrens. Palatka.
	1413	Commelyna erecta. Lin.	Commelynaceae.	Sandy swamps. Palatka.
	1414	Cyperus Michauxianus.Schultes	Cyperaceae.	Swamps and ditches.
	1415	" tetragonus. Ell.	"	Dry sandy soil along the coast.
	1416	" Grayii. Torr.	"	Dry sandy pine barrens.
	1417	" ovularis. Torr.	"	Wet soil.
	1418	Kyllingia sesquiflora. Torr.	"	Low, exposed places. Mid. Fla.
	1419	Dulichium spathaceum. Rich.	"	Ponds and ditches.
	1420	Hemicarpha subsquarrosa. Nees.	"	Low sandy places.
"	1421	Eleocharis cellulosa. Torr.		Marshes. West Florida.
"	1422	Scirpus maritimus. Lin.		Muddy banks. Chipola.
"	1423	Fimbristylis laxa. Vahl.		Low grounds and waste places.
"	1424	" congesta. Torr.		Banks of Apalachicola river.
"	1425	Isolepsis ciliatifolia. Torr.		Dry sandy soil.
"	1426	" stenophylla. Torr.		Dry sandy soil.
"	1427	Rhynchospora distans. Nutt.		Low pine barrens.
"	1428	" trichoides.		Open pine barrens.
"	1429	" alba. Vahl.		Wet springy places.
"	1430	Dichromena leucocephala. Michx.		Damp soil.
"	1431	Scleria reticularis. Michx.	"	Margins of ponds.
"	1432	Muhlenbergia diffusa. Schreb.	Gramineae.	Shaded waste places.
"	1433	" capillaris. Kunth.	"	Sandy soil along the coast.
"	1434	Calamagrostis coarctata. Torr.	"	Fernandina. East Florida.
"	1435	Aristida virgata.	"	Dry soil.
"	1436	" " var. palustris.	"	Margins pine barren ponds.
"	1437	" spiciformis. Ell.	"	Low pine barrens.
"	1438	Spartina polystachia. Willd.	"	Brackish marshes.
"	1439	" glabra. Muhl.	"	Salt marshes.
"	1440	Eustachys glauca. n. sp.	"	Brackish marshes. West Fla.
"	1441	Leptochloa mucronata. Muhl.	"	Cultivated fields.
"	1442	" dubia. Nees.	"	South Florida.
"	1443	Tricuspis sessderioides. Torr.	"	Dry soil.
"	1444	Triplasis Americana. Beauv.	"	Dry sandy soil.
"	1445	Brizopyrum spicatum. Hook.	"	Low sandy shores and marshes.
"	1446	Eragrostis reptans. Nees.	"	Low sandy places.
"	1447	" conferta. Trin.	"	River banks.
"	1448	" tennis. Gray.	"	River banks.
"	1449	" capillaris. Nees.	"	Dry uncultivated fields.
"	1450	" nitida.	"	Low grassy places.
"	1451	" pectinata. Gray.	"	Dry sterile soil.
"	1452	" " var. refracta	"	Damp soil.
"	1453	Paspalum distichum. Lin.	"	Swamps and low grounds.
"	1454	" vaginatum. Swartz.	"	Saline swamps.

MONTH.	No.	NAME OF PLANT.	NATURAL ORDER.	LOCALITY.
Aug. & Sept.	1455	Paspalum Floridanum. Michx.	Gramineae.	Damp soil near the coast.
"	"	1456	" racemulosum. Nutt.	"	Dry sandy soil.
"	"	1457	Panicum filiforme. Lin.	"	Dry sandy soil.
"	"	1458	" anceps. Lin.	"	Damp sterile soil.
"	"	1459	" virgatum. Lin.	"	Sandy soil.
"	"	1460	" Crusgalli. Lin.	"	Wet places.
"	"	1461	Manisuris granularis. Swartz.	"	Fields and pastures.
"	"	1462	Andropogon oligostachys.	"	Dry sand ridges. Middle Fla.
"	"	1463	" scoparius. Michx.	"	Dry sterile soil.
"	"	1464	Rhynchosia Caribaea. D. C.	Leguminosae.	St. Lucy, South Florida.
"	"	1465	Coreopsis aurea. Ait.	Compositae.	Swamps. Palatka's vicinity.
"	"	1466	Spilanthes Nuttallii. Torr. & G.	"	Inundated places.
"	"	1467	Verbena urticifolia. Lin.	Verbenaceae.	Dry, light soil.
"	"	1468	Lycopus sinuatus. Ell.	Labiatae.	Ponds and ditches.
"	"	1469	" " var. intermedius.	"	Ponds and ditches.
	1470	Lycopus sinuatus. var. angustifolius. Benth.		Ponds and ditches.
"	"	1471	Monarda punctata.	"	Dry sandy soil. Palatka.
"	"	1472	Ipomoea Pes caprae. Sweet.	Convolvulaceae.	Drifting sands along the coast.
"	"	1473	" commutata. R. & S.	"	Margins swamps; cult. gr'nds.
"	"	1474	" triloba. Lin.	"	Fort Myers. South Florida.
"	"	1475	" pandurata. Meyer.	"	Riv'r b'nks; m'rgins sw'ps. Pal.
"	"	1476	" " var. hastata	"	Sandy pine barrens.
"	"	1477	" Bona nox. Lin.	"	Salt marshes. Palatka.
"	"	1478	Cuscuta Gronovii. Willd.	"	L'w sh'dy pl'ces; on sm'll h'rbs.
"	"	1479	Acnida australis. Gray.	Amarantaceae.	Marshes and river banks.
"	"	1480	Euphorbia discoidalis. n. sp.	Euphorbiaceae.	Dry, sandy pine b'rr'ns n'ar c'st.
"	"	1481	Fimbristylis spadicea. Vahl.	Cyperaceae.	Salt marshes.
"	"	1482	Scleria laxa. Torr.	"	Damp pine barrens.
"	"	1483	Triplaris purpurea.	Gramineae.	Drifting sand along the coast.
September.	1484	Warea amplexifolia. Nutt.	Cruciferae.	Sand hills.
"	1485	" cuneifolia. Nutt.	"	Sand hills.
"	1486	Desmodium molle. D. C.	Leguminosae.	Waste places. Middle Florida.
"	1487	Archangelica dentata.	Umbelliferae.	Dry pine barrens.
"	1488	Vernonia fasciculata. Michx.	Compositae.	Low grounds.
"	1489	Carphephorus pseudo Liatris. Cass.	"	Open grassy pine barrens. WFl
	1490	Carphephorus corymbosus. Torr. & Gray.		Margins of swamps.
"	1491	Liatris tenuifolia. Nutt.		Dry pine barrens.
"	1492	" gracilis. Pursh.		Sandy pine barrens.
"	1493	" graminifolia. Walt.		Light dry soil.
"	1494	" scariosa. Willd.		Light dry soil.
"	1495	" odoratissima. Willd.		Flat pine barrens.
"	1496	" fruticosa. Nutt.		East Florida.
"	1497	Kuhnia eupatorioides. Lin.		Light and dry soil.
"	1498	Eupatorium hyssopifolium. Lin.		Low grounds.
"	1499	" cuneifolium. Willd.		Rich Shaded soil.
"	1500	" leucolepis. Torr. & Gray.		Flat pine barrens.
"	1501	Eupatorium parviflorum. Ell.		Margins pond and wet places.
"	1502	" teucrifolium. Willd.		Damp soil.
"	1503	" album. Lin.		Dry sandy soil.
"	1504	" mikanioides. n. sp.		St. Vincent Island. West Fla.
"	1505	" pinnatifidum. Ell.		Dry soil. Palatka.

MONTH.	No.	NAME OF PLANT.	NATURAL ORDER	LOCALITY.
September.	1506	Eupatorium perfoliatum. Lin.	Compositae.	Low grounds. Palatka.
"	1507	" serotinum. Michx.	"	Rich soil.
"	1508	" villosum. Swartz.	"	Fort Myers. South Florida.
"	1509	" ageratoides. Lin.	"	Rich, shaded woods. Palatka.
"	1510	" incarnatum. Walt.	"	Rich shaded soil.
"	1511	Conoclinium coelestinum. D.C.	"	Rich soil.
"	1512	Aster divaricatus. Nutt.	"	Salt marshes. Palatka
"	1513	Diplopappus linariifolius. Hook		Dry open woods. West Florida.
"	1514	" amygdalinus. Torr. & Gray.		Swamps in suburbs of Palatka.
"	1515	Solidago discoidea. Torr.& Gr.		Rich woods.
"	1516	" caesia. Lin.		Damp shady woods and banks.
"	1517	" flavovirens. n. sp.		Brackish marshes. West Fla.
"	1518	" virgata. Michx.		Pine barren swamps. Palatka.
"	1519	" sempervirens. Lin.		Salt marshes.
"	1520	" Bootii. Hook.		Sandy soil.
"	1521	" altissima. Lin.		Low thickets. Palatka.
"	1522	" pilosa. Walt.		Low grounds.
"	1523	" tortifolia. Ell.		Dry sandy soil.
"	1524	" brachyphylla. Chapm.		Dry light soil.
"	1525	Bigelovia nudata. D. C.		Low pine barrens.
"	1526	Isopappus divaricatus. T.&G.		Sandy fields and woods.
"	1527	Chrysopsis graminifolia. Nutt.		Sandy pine barrens.
"	1528	" Mariana. Nutt.		Sandy pine barrens.
"	1529	" trichophylla. Nutt.		Dry pine barrens.
"	1530	" gossypina. Nutt.		Dry sandy soil.
"	1531	Pluchea bifrons. D. C.		Margins of pine barren ponds.
"	1532	" foetida. D. C.		Damp soil.
"	1533	" camphorata. D. C.		Salt marshes.
"	1534	" pycnostachyum. Lin.		Swamps and low grounds.
"	1535	Eclipta longifolia. Schrad.		West Florida. Wet places.
"	1536	Helianthus tomentosus. Michx.		Open woods. Palatka.
"	1537	" doronicoides. Lam.		Fort Myers. South Florida.
"	1538	" strumosus. Lin.		Dry soil. Common.
"	1539	" divaricatus. Lin.		Dry woods.
"	1540	" microcephalus. Torr. & Gray.		Dry woods.
"	1541	Actinomeris squarrosa. Nutt.		River banks. Palatka.
"	1542	Verbesina Virginica. Lin.		Dry open woods.
"	1543	Flaveria linearis. Lagarca.		Key West and Southern districts
"	1544	Helenium tenuifolium. Nutt.		Waste places. West Florida.
"	1545	Baldwinia uniflora. Ell.		Low pine barrens. Palatka.
"	1546	Actinospermum angustifolium.	"	Dry sandy ridges in pine barrens
"	1547	Artemisia caudata. Michx.	"	Dry open woods. West Florida.
"	1548	Nabalus Fraseri. D. C.	"	Dry sterile soil.
"	1549	" virgatus. D. C.	"	Damp soil. Suburbs of Palatka.
"	1550	Mulgedium acuminatum. D.C.	"	Margins of fields.
"	1551	Utricularia striata. Leconte.	Lentibulaceae.	Still water.
"	1552	" bipartita. Ell.		Miry marg. ponds West Florida.
"	1553	Scrophularia nodosa. Lin.	Scrophulariaceae.	Shaded banks; thickets. Palatka
"	1554	Chelone glabra. Lin.	"	Wet banks of streams.
"	1555	Gerardia linifolia. Nutt.	"	Low pine barrens. Palatka.
"	1556	" divaricata. n. sp.	"	Low sandy pine bar'ns. W. Fla.
"	1557	" filicaulis. Benth.	"	Low grassy pine barrens. W. Fla
"	1558	" aphylla. Nutt.	"	Low sandy pine barrens.
"	1559	" purpurea. Lin.	"	Low grounds. Palatka.

MONTH.	No.	NAME OF PLANT	NATURAL ORDER.	LOCALITY.
September.	1560	Gerardia maritima. Raf.	Scrophulariaceae.	Salt marshes.
"	1561	" setacea. Ell.	"	Dry sandy pine barrens.
"	1562	" tenuifolia. Vahl.	"	Light soil.
"	1563	" filifolia. Nutt.	"	Low sandy pine barrens.
"	1564	" parvifolia. Benth.	"	Grassy margins ponds.
"	1565	Collinsonia Canadensis. Lin.	Labiatae.	Rich shaded woods.
"	1566	" scabriuscula. Ait.	"	Rich shaded woods.
"	1567	" punctata. Ell.	"	Rich shady woods.
"	1568	Lycopus Virginicus. Lin.	"	Ponds and ditches. Palatka.
"	1569	Amblogyna polygonoides. Raff.	Amarantaceae.	Tampa. South Florida.
"	1570	Scleropus crassipes. Moquin.	"	South Florida.
"	1571	Iresine vermicularis. Moquin.	"	Sandy sea shore.
"	1572	" diffusa. H. & B.	"	Saline marshes. South Florida
"	1573	Polygonella brachystachya.	Polygonaceae.	Fort Myers. South Florida.
"	1574	Cyperus vegetus. Willd.	Cyperaceae.	Low pine barrens. Palatka.
"	1575	" Drummondii. Torr.	"	Shady swamps. Middle Florida.
"	1576	Scirpus leptolepis.	"	Lakes and ponds. Middle Florida
"	1577	Isolepis Warei. Torr.	"	West Fla. Dry sands near coast.
"	1578	Sporobolus Floridanus. n. sp.	Gramineae.	Low pine barrens. Mid. & W. Fla
"	1579	Muhlenbergia trichopodes.	"	Low pine barrens.
"	1580	Leptochloa polystachya. Kunth.	"	Brackish swamps along coast.
"	1581	Phragmitis communis. Trin.	"	Deep river marshes near coast.
"	1582	Paspalum undulatum. Poir.	"	Low cultivated grounds.
"	1583	" Blodgettii. n. sp.	"	Key West and southern districts
"	1584	Panicum gymnocarpum. Ell.	"	Muddy banks of rivers. Palatka
"	1585	" amarum. Ell.	"	Drifting sands along the coast.
"	1586	" proliferum. Lam.	"	Wet places near the coast.
"	1587	" capillare. Lin.	"	Sandy fields. Common.
"	1588	" verrucosum. Muhl.	"	Swamps.
"	1589	" angustifolium. Ell.	"	Dry soil.
"	1590	" rufum. Kunth.	"	Pine barren swamps.
"	1591	Rottboellia rugosa. Nutt.	"	Pine barren ponds and swamps
"	1592	Andropogon tener. Kunth.	"	Dry grassy pine barrens.
"	1593	" Nuttallii.	"	Low pine barrens.
"	1594	" furcatus. Muhl.	"	Open woods.
"	1595	" tetrastachyus. Ell.		Low pine barrens.
"	1596	" macrourus. Michx.		Low pine barrens.
"	1597	" melanocarpus. Ell.		Old fields. Indian river.
"	1598	Erianthus strictus. Baldw.		Dry, sometimes wet soil.
"	1599	Sorghum avenaceum.		Dry sandy soil.
"	1600	" nutans. Gray.		Dry barren soil.
Sept. to Oct.	1601	Petalostemon corymbosum. Michx.	Leguminosae.	Dry pine barrens.
"	1602	Liatris paniculata. Willd.	Compositae.	Damp pine barrens.
"	1603	Eupatorium foeniculaceum. Willd.	"	Old fields around Palatka.
"	1604	Eupatorium coronopifolium.		Dry sandy soil.
"	1605	Aster spectabilis. Ait.		Pine barrens.
"	1606	" paludosa. Ait.		Low pine barrens.
"	1607	" concolor. Lin.		Dry sandy soil.
"	1608	" squarrosus. Walt.		Dry soil. Palatka.
"	1609	" coccineus. Willd.		Southern districts.
"	1610	" asperulus. Torr. & Gray.		Dry gravelly soil. W. Florida.
"	1611	" sagittifolius. Willd.		Rich woods.
"	1612	" Baldwinii. Torr. & Gray.		Dry pine barrens.
"	1613	" dumosus. Lin.		Dry soil. Common.

MONTH	No.	NAME OF PLANT.	NATURAL ORDER.	LOCALITY.
Sept. to Oct.	1644	Aster simplex. Willd.	Compositae.	Low grounds.
" "	1645	Boltonia diffusa. Ell.	"	Damp soil.
"	1646	Solidago pulverulenta. Nutt.	"	Damp pine barrens.
"	1647	" speciosa. Nutt.	"	Dry soil.
"	1648	Chrysopsis scabrella. Torr. & Gray.	"	Pine woods around Palatka.
	1649	Baccharis halimifolia. Lin.	"	Low grounds near the coast.
	1650	Eclipta erecta. Lin.	"	Wet places.
	1651	Helianthus angustifolius. Lin.	"	Low grounds.
	1652	" antrorubens. Lin.	"	Dry soil.
	1653	Coreopsis gladiata. Walt.	"	Low pine barrens.
	1654	" angustifolia. Ait.	"	Pine barren swamps.
	1655	Bidens chrysanthemoides. Mchx	"	Wet places in vicinity of Palatka
	1656	Spilanthes repens. Michx.	"	Muddy banks.
	1657	Cacalia suaveolens. Lin.	"	Low grounds. West Florida.
	1658	Cosmos caudatus. Kunth.	"	Key West and Southern districts
	1659	Hieracium Gronovii. Lin.	"	Dry sandy soil.
	1660	Lobelia amoena. Michx.	Lobeliaceae.	Swamps around Palatka.
	1661	Gentiana ochroleuca. Froel.	Gentianaceae.	Dry sandy woods.
	1662	Polygonum densiflorum. Meisner.	Polygonaceae.	Muddy banks. Palatka.
	1663	Thysanella fimbricata. Gray.	"	Dry pine barrens.
	1664	Coccoloba uviferae. Jacq.	"	Indian river. South Florida.
	1665	Ceratophyllum demersum. Lin	Ceratophyllaceae.	In still water. Rice cr'k; Palatka
	1666	Apteria setacea. Nutt.	Burmanniaceae.	Grassy or mossy marg. swamps.
	1667	Ponthieva glandulosa. R. Br.	Orchidaceae.	Low shady woods. Palatka.
	1668	Thalia divaricata. n. sp.	Cannaceae.	Ponds. West and South Fla.
	1669	Tofieldia pubens. Ait.	Melanthaceae.	Low pine barrens about Palatka
	1640	Xyris serotina. n. sp.	Xyridaceae.	Pine barren swamps. W. Fla.
	1641	Gymnopogon racemosus. Beauv.	Gramineae.	Dry sandy soil. Palatka.
	1642	Paspalum fluitans. Kunth.		River swamps.
	1643	Amphicarpum Purshii. Kunth.		Banks of the Apalachicola.
	1644	Rottboellia corrugata. Baldw.		Low pine barrens near coast.
	1645	Andropogon argenteus. Ell.		Wet pine barrens about Palatka
	1646	Erianthus alopecuroides. Ell.		Dry or wet soil.
	1647	" contortus. Ell.		Dry or wet soil.
	1648	" brevibarbis. Michx.		Dry or wet soil.
	1649	Sorghum secundum. Chapm.		Dry sandy ridges in pine barrens
	1650	Monanthochloë littoralis. Engelm.		Low sandy shores. South Fla.
Sept. to Nov.	1651	Calamintha dentata. n. sp.	Labiatae.	Sandy ridges. W. & S. W. Fla.
" "	1652	Burmannia biflora. Lin.	Burmanniaceae.	Margins of ponds and swamps.
" "	1653	Tripterella capitata. Michx.	"	Swampy pine barrens.
October.	1654	Aster Chapmannii. Torr. & Gray.	Compositae.	Pine barrens swamps. W. Fla.
	1655	Aster linifolius. Lin.		Wet places along coast.
	1656	Solidago gracillima. Torr. & Gray.		Dry pine barrens. Palatka subs.
	1657	Solidago odora. Ait.		Dry soil.
	1658	" amplexicaulis. Torr. & Gray.		Dry open woods.
	1659	Solidago Laevenworthii. Torr. & Gray.		Damp soil.
	1660	Solidago Canadensis. Lin.	"	Margins of fields. Palatka.
	1661	" serotina. Ait.	"	Low grounds.

MONTH.	No.	NAME OF PLANT.	NATURAL ORDER.	LOCALITY.
September.	1662	Solidago pauciflosculosa. Michx.	Compositae.	Sandy banks and shores.
	1663	Baccharis angustifolia. Michx.		Saline marshes.
	1664	Helianthus Radula. Torr. & Gray.		Low sandy pine bar'ns. Palatka
··	1665	Helianthus heterophyllus.	··	Pine barren swamps.
··	1666	Lobelia brevifolia. Nutt.	Lobeliaceae.	Damp open pine bar'ns. Palatka
··	1667	·· glandulosa. Walt.	··	Pine barren swamps. Palatka.
··	1668	Gentiana Elliottii. Chapm.	Gentianaceae.	Middle to South Florida.
··	1669	·· Ell. var. latifolia.	··	Pine barren swamps near coast.
··	1670	·· Ell. var. parvifolia.	··	River banks. Middle Florida.
··	1671	Spiranthes cernua. Richard.	Orchidaceae.	Grassy swamps near Palatka.
··	1672	·· odorata. Willd.	··	Muddy banks near Marianna.
··	1673	Pleea tenuifolia. Michx.	Melanthaceae.	Pine barren swamps.
··	1674	Lachnocaulon glabrum. Kornick.	Eriocaulonaceae.	Sandy spring places. Palatka.
··	1675	Scleria filiformis. Swartz.	Cyperaceae.	St. Lucy. South Florida.
	1676	Leptochlea Domingensis. Link	Gramineae.	Tampa. South Florida.
Oct. to Nov.	1677	Parnassia Caroliniana. Michx	Parnassiaceae.	Dampsoil. Palatka to Rice creek
·· ··	1678	Sageretia Michauxii. Brongm	Rhamnaceae.	Dry sandy soil along the coast.
·· ··	1679	Aster adnatus. Nutt.	Compositae.	Sandy barrens.
·· ··	1680	·· Elliottii. Torr & Gray.	··	Swamps. Common.
··	1681	·· Carolinianus. Walt.	··	River swamps. Palatka.
··	1682	Chrysopsis decumbens. n. sp.	··	Shores St. Vincent Island.
··	1683	Calamintha coccinea. Benth.	Labiateae.	Shores St. Andrews Bay. S. Fla.
··	1684	Dicerandra linearifolia. Benth	··	Dry sandy pine barrens.
··	1685	·· densiflora. Benth.	··	East to Middle Florida.
··	1686	Spiranthes brevifolia. n. sp.	Orchidaceae.	Open woods in the pine barrens.
Oct. to Dec.	1687	Solanum verbascifolium. Lin.	Solanaceae.	Indian river. South Florida.
November.	1688	Centrosema Plumiera. Turp.	Leguminosae.	Indian river. South Florida.
··	1689	Galactia filiformis. Benth.	··	Keys of South Florida.
··	1690	Ecastaphyllum Brownei. Pers.		Indian and Manatee rivs. S. Fla.
··	1691	Hamamelis Virginica. Lin.	Hamameliaceae.	Low woods around Palatka.
··	1692	Archemora ternata. Nutt.	Umbelliferae.	Low swampy pine barrens.
··	1693	Capraria biflora. Lin.	Scrophulariaceae.	Indian river. St. Lucy. S. Fla.
··	1694	Pilea herniarioides. Link.	Urticaceae.	Shaded moist places. Key West
··	1695	Cyperus confertus. Swartz.	Cyperaceae.	Fort Myers. South Florida.
··	1696	·· filiformis.	··	Key West and westward, S. Fla.
Nov. to Dec.	1697	Bidens leucantha. Willd.	Compositae.	Brooksville. South Florida.
·· ··	1698	Vernonia pumila. Chapm.		Punta Gorda. South Florida.
·· ··	1699	Gentiana angustifolia. Michx.	Gentianaceae.	Low pine barrens.
December.	1700	Borrichia arborescens. D. C.	Compositae.	Dry sand along the coast.

www.ingramcontent.com/pod-product-compliance
Lightning Source LLC
Chambersburg PA
CBHW021450090426
42739CB00009B/1696